DEEP INTO DECO

The Diver's Decompression Textbook

SECOND EDITION

Asser Salama

Copyright © 2015, 2018 by Best Publishing Company

International Standard Book Number: 978-1-947239-09-8

Library of Congress catalog card number: 2018935741

Best Publishing Company
631 US Highway 1, Suite 307
North Palm Beach, FL 33408

Cover Photo courtesy of René Andersen.

Dedicated to my mother

CONTENTS

7 OTHER DECOMPRESSION MODELS   61

8 VARIOUS TOPICS   72

FOREWORD - SECOND EDITION

In my nearly half a century of involvement in diver education, the sport and the science of scuba diving has transformed from strictly an interesting and unique pastime to an essential tool in scientific and military diving. Along with this transformation, divers have pushed the boundaries of the recreational diving envelope by applying the latest skills, knowledge, and technology to go beyond traditional recreational diving limits.

In 1981, I had the privilege of being involved in the development of the first electronic decompression computer, the E.D.G.E., which is an acronym for Electronic Dive GuidE. This device revolutionized decompression planning and led to the development of the plethora of decompression computer currently on the market and in use throughout the diving world.

The introduction of an electronic device utilizing a specialized algorithm for calculating a diver's decompression obligation and then providing a pathway for the diver to safely return to the surface created an ethical dilemma in diver education that had, until now, taught divers to utilize decompression tables. The dilemma had to do with using an electronic device to manage a diver's decompression obligation and what to do if that piece of equipment failed. Would the diver understand enough about decompression theory and practice to plan a safe return to the surface?

Having been president of the largest diving safety organization in the world, Divers Alert Network (DAN), I witnessed firsthand the improper application of diving techniques and knowledge resulting in diving injuries and fatalities.

As divers push the boundaries of traditional recreational diving limits, the value of a comprehensive yet immensely readable source of information has become paramount. Asser Salama's *Deep Into Deco: The Diver's Decompression Textbook Second Edition* is an outstanding compendium of information on decompression history, theory, and practice. It also contains supporting information from many of the diving world's most knowledgeable pioneers and innovators.

In analyzing recreational diving accidents reported to DAN, findings indicate the root cause in the vast majority of these diving accidents was sometimes due to a diver making inappropriate decisions when confronted with an emergency situation. Some of these accidents resulted in decompression sickness.

When lecturing as a global diving industry consultant, I begin safety seminars by saying, *"Diving is inherently safe but can be mercilessly unforgiving of mistakes."* Knowledge is a key component to preventing mistakes in diving. Critically important information regarding decompression theory can provide you with valuable tools and insight that can be drawn from when the outcome from those decisions can literally mean the difference between life, death, or serious injury.

Safe divers overcome challenges in and under the water not by physical strength but by the effective application of knowledge and skill.

If you are looking for a textbook that provides a thorough understanding of decompression theory and application and how it might be impacting your safety and the safety of others you dive with, this book is a must read.

— Dan Orr, President, Dan Orr Consulting LLC
President Emeritus, Divers Alert Network (DAN)
Member, Diving Industry and International Scuba Diving
Halls of Fame and NAUI's Hall of Honor
IANTD Nitrox Instructor #10

FOREWORD - FIRST EDITION

This book is long overdue — and it's worth the wait. My professional diving career began in January 1971, working on extreme deep-diving projects for the U.S. Navy, filming fast-attack submarines in the abyss of the Virgin Islands Trench. Back then our only resource to plan such dives was the Navy Exceptional Exposure Tables. They worked, for the most part, but our decompression stops at times seemed longer than high school, and we eagerly sought alternatives. At the time, diving-deco-model physiologists jumped at the chance to have us "beta test" new evolutions in table methodology, using deep gas switches during ascent to high-percentage oxygen in early nitrox mixes and far deeper use of pure oxygen for our primary hangs. This was considered by many to be heresy at the time, but it profoundly increased our efficiency and got us out of the water faster, lessening the time pelagic sharks could try to nibble on us during the longer deco stops.

Later, I was introduced to my dear friend Dr. Bill Hamilton, who pioneered the modern science of decompression modeling and algorithms; he brought the whole subject into the proper perspective. He sagely noted that there would always be innovations and changes, but the most important thing to remember is that "What works... is what actually works!" He also noted that the only way to validate the hypothesis was to go out and try it. I think that's why he secretly loved the idea that he could create a new version of an algorithm and guys like me, Sheck Exley, Jim Bowden, Wes Skiles, Parker Turner, Billy Deans, Tom Mount, and others, would dive it and report back.

It was a unique symbiotic partnership that moved the technology forward quickly and paved the way for the development of the early dive computers. Of course, it was a bit unnerving sometimes after a particularly challenging deep dive to call Bill with a report and then to detect a bit of surprise in his voice that we were still alive.

As CEO of Uwatec, I was privileged later to work with the amazing Swiss physiologist Dr. Albert Bühlmann as we adapted his deco models to be the primary software running our dive computers. This led the way into the modern era and changed diving forever.

What Asser Salama has accomplished with this book is remarkable. He has taken that early history of experimental trial and error and produced a stunning reference text that brings the science into sharp focus. *Deep Into Deco* is not a light read. You are not taking this book to browse through at the beach in the summer, and you are probably not going to read it to pass the time on a long airline flight.

But you will read it to get the single best education and information on decompression modeling that has yet to be produced. And you will come away from it perhaps a bit mentally exhausted but intellectually informed and ready to make the proper decisions when planning responsible dive exposures.

Take a deep breath and dig in. You are going to be here awhile.

— *Bret Gilliam*
March 2014

ABOUT THE AUTHOR

Asser Salama is a technical diver and a decompression modeler, most notably famous as the founder of *Tech Diving Mag* and the developer of Ultimate Planner. Asser has a bachelor's degree in engineering and a master's degree in business administration. For almost a decade, Asser has been diving deep into decompression algorithms in an attempt to implement computational models based on credible research papers in order to prevent some pioneering work from fading into academic obscurity.

Tech Diving Mag is an online technical diving publication aimed at research, development, and exploration. Available free and based on article contribution, the editorials are generally written by tech diving enthusiasts rather than journalists. This is the most suitable place for Asser to run his research-related editorials. He is also the magazine's editor and publisher.

Based on his extensive research, he developed Ultimate Planner, a full-fledged VPM-B and Buhlmann-GF decompression planning tool for OC and CCR divers, with the unique ability of accelerating the no-fly time by breathing rich mixes on the surface. This distinctive piece of software also has the capability of incorporating asymmetric gas kinetics, which adds a second dimension of conservatism for deeper, longer, and/or colder dives.

Later on, Ultimate Planner was equipped with more bells and whistles, like the ability of modeling the inner ear as either aqueous (biological fluid close to water) or lipid (fat) tissue (for ICD prediction).

INTRODUCTION

It was about noon when I went for a solo shore dive on a nonredundant single 80-cubic-foot (11- but wrongfully perceived as 12-liter) air tank and a Suunto dive computer adjusted to the most "aggressive" settings. My surface interval before the dive was more than 24 hours.

While in the water, I decided to get to the deepest part of the reef table. I knew it was at 43 meters (140 feet). The reef here was not an exact wall; it was slopey with some scattered patches. I reached the deepest point in about four minutes. Then, and only then, I opted to go on with the gentle sand slope, away from the main reef. I'd explored this place twice before. The first time I went to 48 meters (160 feet); the second time I explored farther, found two reef patches and had a look at the first one at 61 meters (200 feet). Now it felt like about the right time to take a glimpse at the second patch.

The second patch was at a depth of about 67 meters (220 feet) and far enough from the main reef. That's what I found out. It took me another four minutes to get there, and I had my time "smelling the roses." I kept exploring, and my Suunto moaned, blinked, and amassed sky-rocketing decompression penalties. Jeez, I thought, it's too bad I don't have the good old Uwatec today; it usually doesn't bother me with this crap. I ignored the dive computer and continued enjoying the isolated reef patch I had for myself, until my pressure gauge read 70 bar (1,000 psi). That was certainly most unexpected. Now what?

I didn't like surfacing in the blue, not to mention long surface swims. It was a shore dive, so there was no boat to pick me up. On my way back to the main reef, I decided not to be silly, so I didn't stick to the sand slope. I ascended to the 25-meter (80-foot) mark and continued swimming in midwater. Before reaching the reef table, I started a slow ascent and finally reached the 6-meter (20-foot) mark. I stopped there for a while then ascended to 3 meters (10 feet) and stopped there until my pressure gauge read 10 bar (150 psi). The last thing my Suunto displayed before going into the error mode was 28 minutes of missed decompression "obligation."

I had oxygen onsite, but I didn't use it. I didn't think I needed it. On my way back to the dive center, I drank some water. After rinsing the gear, I took a cold shower (it was almost November and still uncomfortably hot) and then took a short nap. In about 16 hours I was driving my beloved Fiat Siena, heading to Sharm El Sheikh 100 kilometers (60 miles) south of Dahab. The road's highest altitude is 640 meters (2,100 feet). I was going for a three-day boat trip; it was a very nice one indeed, in part because my Suunto was "bent" (running as a bottom timer not a dive computer), so it didn't bother me anymore.

Focusing only on the decompression part and discarding every other aspect of this dive, did I get lucky? Should I have been injured? How was this decompression penalty calculated? How could one twist the schedule in such an "irresponsible" manner and get away with it? More important, why did I whine about not having the Uwatec dive computer that day? Wasn't the Suunto good enough?

On the other hand, if I followed the dive forums on the Internet, I frequently saw the term "undeserved hit," meaning that a diver got "bent" — developed decompression sickness (DCS) — although he or she did everything possible to avoid this unfortunate event. I remember one time a vacation diver got bent while breathing what some operators incorrectly call air28 (28% oxygen, 72% nitrogen), although he used a dive computer (which was adjusted to normal air settings to offer a higher margin of safety) and did not violate the ascent rate or any no-decompression limits (NDL).

Wasn't it enough to stick to what the dive computer indicated? Even diving more conservatively doesn't guarantee you will avoid getting hit. At the same time, some people ignore the dive computer and do whatever they like without getting hit. This seems a bit illogical, doesn't it? What's the advantage of getting a dive computer or even planning the dive using a decompression planning program or a set of dive tables?

The answer: Decompression is still far from being an exact science. There is a severe lack of funding for decompression research, which means whenever research on a particular aspect is done, it might not meet certain criteria in terms of the number of participants, the experimental conditions, etc. The result is many different theories on the very same aspect. These theories sometimes contradict each other; they can't all be valid. Acceptance of what the scientific community believes is another issue. Generally speaking, the diving community applies what works for it rather than what the scientific community subscribes to.

In conclusion, blindly subscribing to your favorite decompression planning tool, whatever it is, and following the most recent industry standards without understanding the underlying principles of decompression and without having enough knowledge on various aspects and theories could prove fatal. This does not mean you should not keep up with the latest standards. It simply means you should learn and understand.

In addition to being a technical diver and instructor, I am an engineer and a software developer. That's why I'm particularly interested in mathematical models. Since 2010, I've been studying decompression algorithms and contacting industry leaders and researchers in an attempt to enhance what we already have in hand. The output is Ultimate Planner, a decompression planning tool with some unique features. While reading this book, you'll see some examples of these features, how they were developed and how to use them. You can download a Lite version of Ultimate Planner at www.techdivingmag.com/ultimateplanner.html.

This book will touch on the basic principles of decompression theory and at the same time shed light on the latest developments and controversial issues. You'll read some interesting interviews with researchers, accomplished divers, industry professionals, and software developers. I've also quoted experts on historical perspectives and other more specific issues, so you'll find the style a mix of strict no-nonsense writing and interesting storytelling. I didn't use footnotes, but references are numbered in the text and collected at the end of the book.

— Asser Salama
2018

CHAPTER 1
Historical Perspective

History often strikes a nerve with many people, particularly those who seem to appreciate the look at how things strike up. Some storytelling is in order prior to starting. Dr. Michael Powell, the renowned decompression physiology researcher at NASA (retired), narrates:

> *"Bubbles in living organisms were first discovered by Robert Boyle (1627–1691) in 1660. In addition to discovering 'decompression bubbles,' Boyle was a polymath. He excelled at alchemy [writing* The Sceptical Chymist*], physics, philosophy and theology. Divers are familiar with that name as it relates to Boyle's Law that relates pressure and volume of gasses. It is this law that can kill you if you ascend with your breath held. Serious business.*

> *"Using an air pump invented by his young assistant Robert Hooke, Boyle observed the effects of rarified air pressure on live plants and animals. He wrote, 'I once observed a viper furiously tortured in our exhausted receiver … that had manifestly a conspicuous bubble moving to and fro in the waterish humor of one of its eyes.' Thus, the first bubbles in living creatures were actually in altitude depressurization."* [1]

A couple of centuries later, it was noticed that people working at elevated pressures in dry underwater pressurized boxes (caissons) or construction tunnels beneath rivers suffered muscle cramping, pain, and symptoms of paralysis. These symptoms were referred to as caisson disease. Decades later they were also known as the "bends" because the affected workers often had trouble standing straight, so they took bowed positions to relieve the pain. The workers themselves were called caisson workers or simply caissons.

By this time, French scientist Paul Bert (1833–1886), also known as the "father of aviation medicine," was studying the physiological effects of air pressure, both below and above sea level. His comprehensive investigation into this matter was described in detail in his 1878 classical piece of work *La Pression Barométrique*. Bert was the first to show that oxygen can be toxic under pressure. Central nervous system (CNS) oxygen toxicity was first described in his publications; to this day many still call it the Paul Bert effect. He was also the first to observe evident bradycardia (slowness of the heartbeat) in ducks while diving. This laid the foundation to an advanced diving technique we know now as the mammalian reflex.

Paul Bert
Photo courtesy of Wikipedia.
Public Domain. "Haldane's decompression model."

Bert conducted some studies on dogs. By exposing them to pressure, he was able to identify the gas causing the problem: nitrogen. The problem was further recognized as bubbles formed in the tissues and the blood. Bert suggested that caisson disease could be avoided by permitting only a slow decrease in ambient pressure.

"The longer the workmen remain in the caissons, the more slowly they should undergo decompression, for they must allow not only time for the nitrogen of the blood to escape, but also allow the nitrogen of the tissue time to pass into the blood. This last point is the most difficult to obtain."

— Paul Bert, *La Pression Barométrique*, 1878

In an attempt to treat caisson disease, Bert suggested that the affected workers return to the caisson and ascend slowly. He also observed that breathing pure oxygen alleviates the symptoms.

Bert's discoveries paved the road for a quantum leap. Powell continues:

J.S. Haldane
Photo courtesy of Wikipedia.
Public Domain. "Haldane's decompression model.

"Virtually all decompression procedures in use today can be traced to John Scott Haldane (1860–1936), a respiratory physiologist who worked in England from the late 1800s to the early 1900s. His research on bubbles and decompression were among the earliest works. He founded the Journal of Hygiene, *studied poisonous gases in coal mines, problems of heat stroke, chlorine gas in warfare [he made a form of gas mask during World War I], respiration at altitude and, for us, decompression sickness. He was tasked by the Admiralty's Deep Diving Committee to produce effective decompression tables to eliminate 'caisson disease,' as DCS was then known. He produced tables and also performed studies on the physiology of decompression sickness.*

"The developer of the 'stage decompression method' used today, Haldane originated the concept of rising to a depth near the surface in steps as contrasted with the method of slow linear ascents that was employed at that time. This rise in stages is known as the 'Haldane Method,' known to every scuba diver, if not by that name.

"As the other portion of his algorithm [i.e., a calculation method], Haldane reasoned that blood flowed to various organs of the body in varying amounts, and thus the organs gained [or lost] nitrogen at different rates. To handle this exchange, he conceived of tissues with gain and loss expressed as an exponential function. Time was in the exponent, and half filled or half empty yielded a time known as the 'halftimes.' This is the same system that we are familiar with in the decay of radioactive material and the associated radioactive 'halftimes.'"

The halftime system Haldane devised (and still in use today) corresponds with Bert's "most difficult point to obtain." Powell resumes:

"Haldane knew that divers could ascend by a certain number of feet and not get 'caisson disease' if these upward excursions were not too great. He reasoned that if 'bends' did not occur then bubbles did not form if this upward ascent was limited to a short jump. Haldane had seen evidence that bubbles formed easily in supersaturated fluids outside of the body. Something was present there that was absent in the bodies of living animals.

"'The urine found in the bladder postmortem is remarkably free from bubbles; on two occasions only has free gas been found. We have evidence here that the phenomenon

must be due to supersaturation and the absence of 'points,' since we have very frequently observed goats pass urine after decompression which frothed freely on coming into contact with foreign surfaces.'

— *J.S. Haldane. The prevention of compressed air illness.* J. Hygiene Camb. *1908, p. 415*

"He knew that something was needed for bubbles to form in the living body, and he called these 'points.' [We now know them to be the gas 'seeds' or tissue microbubbles.]

"This is well seen on watching under the microscope a stream of bubbles coming off some 'point' in soda water. It follows that if the concentration of dissolved molecules of gas is not higher than some unknown point [we would call this partial pressure today], *bubbles will not be formed. It is possible that the absence of bubbles from most of the solid tissues is to be explained by this non-existence of very small bubbles* [we would call these 'micronuclei' today] *and the mechanical difficulties of the rapid aggregation of a sufficient number of molecules to produce large bubbles."*

— *J.S. Haldane. The prevention of compressed air illness.* J. Hygiene Camb. *1908, p. 422*

"He did not know the details of micronuclei to the level known today — but that is not to say that the story is complete even now."

Haldane conducted extensive studies on decompression, using goats in hyperbaric chambers. He concluded that if no symptoms of DCS were present postdecompression,

Haldane used goats in hyperbaric chambers for his decompression studies.
Photo courtesy of Wikipedia. Public Domain. "Haldane's decompression model."

then no bubbles were formed in the blood. (However, Doppler tests now prove otherwise.) He suggested that three hours are enough for nitrogen loading to reach saturation in goats; he presumed humans would reach full saturation in five hours.

Accordingly, he designed his "slowest" tissue (we now refer to this as "compartment") to almost saturate in five hours. His slowest compartment halftime was 75 minutes. This hypothetical tissue would reach 93.75% of full saturation in 5 hours. His other compartment halftimes were 5, 10, 20 and 40 minutes. The final outcome of Haldane's work was three tables for compressed-air diving: The first was for all dives requiring less than 30 minutes of decompression time, the second was for all dives requiring more than 30 minutes of decompression time, and the third was for deep air diving to 100 meters (330 feet) with oxygen for decompression. These schedules were characterized by a relatively rapid ascent from depth to the initial decompression-stop depth (which he assumed to be half the absolute pressure on the diver) then followed by a noticeably slower ascent rate to the surface. The Royal Navy adopted these tables in 1908 and continued to use them with revisions for about 50 years. In 2014 we still refer to the calculation methods compatible with the Haldane rationale as Haldanean or neo-Haldanean.

On the other side of the Atlantic, the U.S. Navy (USN) produced its first tables[2] in 1915, based on the pioneering work of George Stillson. They were called the C&R tables because they were published by the Bureau of Construction and Repair. These tables were used with success in the salvage operation of the submarine USS F-4 at a depth of 93 meters (305 feet).

During the 1930s, James Hawkins, Charles Shilling, and Raymond Hansen conducted extensive experimental dives. The existing tables were not suitable for long, deep dives. The conclusion was that the allowable supersaturation ratio was not constant; it was a function of the compartment halftime. In 1937 O.D. Yarborough further developed their work and produced a set of tables for the U.S. Navy based

The USS F-4 (circa 1914)
Photo courtesy of Wikipedia. Public Domain.
"USS F-4."

on the 20-, 40- and 75-minute halftime compartments only. The two fastest compartments (5 and 10 minutes) were dropped. These tables were used for two good decades. In 1956 the revised U.S. Navy Standard Air Decompression Tables came into use, with the two fastest compartments restored into the model and a much slower one (120 minutes) introduced.

On the European side, Val Hempleman was taking another turn. In 1952 he introduced the slab diffusion model, a totally different approach than Haldane's.[3] In 1960 he did not subscribe to yet another basic assumption of Haldane's, which was that gas uptake and elimination took identical times. He presented evidence from animal studies that the uptake and elimination of inert gases were not symmetrical and assumed that the elimination process was one and a half times slower than that of the uptake,[4] now referred to as asymmetric gas kinetics. The Royal Navy adopted his work. Known as the Royal Navy Physiological Laboratory (RNPL) tables, they were first published in 1968 and modified in 1972. The British Sub-Aqua Club (BSAC) adopted a version of the 1972 modified tables, using them until 1987, publishing their own set in 1988.

In 1957 Robert Workman added three compartments with 160-, 200- and 240-minute halftimes. He assumed the total number of compartments to be nine. In 1965 he extended his work to include helium and introduced the concept of M-values (M for *maximum*). The concept was simple: Each compartment would have a maximum inert gas tension that could be safely tolerated at any given depth and at the surface without bubble formation.

In 1959 Swiss hyperbaric medicine pioneer Albert Bühlmann (1923–1994) started research on decompression. He was particularly interested in defining the decompression implications of diving at altitude. He also tried to determine the slowest halftime compartment value. His general approach for calculation was similar to Workman's. In 1983 his work was first published in German in a reference book titled *Dekompression: Dekompressionskrankheit* (in English, *Decompression: Decompression Sickness*). It was translated to English one year later. In the early 1990s the title was revised to *Tauchmedizin*, or *Diving Medicine*. Three more editions were published in German, two of which were translated to English. The last edition (1995) is still available only in German. Bühlmann's algorithms came into use in the form of custom tables and were also integrated into popular dive computers such as Uwatec. Several software planning tools incorporate modified versions of Bühlmann's algorithms.

In the 1970s Doppler technology was incorporated into decompression research. This then-new technology allowed researchers to measure the existence of bubbles in the diver's body. The Doppler device did this by sending an ultrasonic signal, receiving its reflection back from the bubbles in the test subject and subsequently emitting a sound. More bubbles meant more emitted sounds. The sound then got scaled to indicate the approximate amount of bubbles. The Defence and Civil Institute of Environmental Medicine (DCIEM, renamed Defence Research and Development Canada (DRDC) in 2002) started to utilize this breakthrough technology in 1979. The DCIEM tables were finally published in 1992. This set gained wide popularity and is regarded as one of the safest tables. After more than two decades, it is still in use.

In 1984 Edward Thalmann (1945–2004) published another algorithm based on the asymmetric gas kinetics concept: the USN E-L algorithm. It assumed nitrogen was absorbed by tissues at an *exponential* rate (as in other Haldanean models) but was discharged at a slower *linear* rate. In 2001 some U.S. Navy divers made the first official computerized decompression dives in U.S. military history, using dive computers made by Cochran Undersea Technology and incorporating the USN E-L algorithm. The Navy Experimental Diving Unit (NEDU) continued research, and in 2007 Wayne Gerth and David Doolette produced a set of tables based on Thalmann's work.

In 1986 David Yount and Don Hoffman introduced a dual-phase decompression model. Their approach was to consider both the dissolved gas and the gas in bubbles in the blood and the body tissues. They produced what we know now as varying permeability model (VPM). The model was completed in 2000, and in 2002 Erik Baker introduced a revision. The 2002 revision compensates for bubble expansion and contraction using Boyle's Law and is known as VPM-B. Several variations of VPM-B are now integrated into some technical-diving computers and are incorporated into several modern software planning tools. The most popular model variations are Ross Hemingway's /E (2005), Shearwater Research's /GFS (2011) and my /U (2011).

Those are just some of the milestones. Scores of papers, tables, models and studies have been published, and there's still a lot of research to be done in virtually every aspect of the decompression theory. We're way too far from being "there."

Summary

In this chapter, we touched on some of the major landmarks in the history of decompression research. The primary emphasis was on the significant discoveries and the evolution of the major theories. The pioneering efforts of Boyle, Bert, Haldane, Hempleman, Workman, Bühlmann, Thalmann, Yount, and Hoffman, along with other researchers and developers were highlighted.

CHAPTER 2

Basic Decompression Principles

At the surface, we breathe "normal" air. Air is a mixture of many gases, mainly nitrogen (N_2, a little bit more than 78%), oxygen (O_2, almost 21%) and argon (Ar, almost 1%). Our body uses the O_2 from the air in a process called metabolism and discharges the waste products of this process, along with the other unused gases, through exhalation. Practically speaking, for scuba-diving applications, the air we breathe consists of 21% O_2 and 79% N_2. Since the body neither uses nor reacts with the N_2 at the surface, we'll call it inert gas.

The term "normal" air means that the air is at its normal pressure at the surface, which is the atmospheric pressure (frequently referred to as barometric pressure). The atmosphere is a layer of gases surrounding our planet and is retained by the Earth's gravity. Atmospheric pressure is the force per unit area exerted by the atmosphere. In layman's terms, atmospheric pressure is the weight of the air column above us that exerts force on our bodies. Under normal conditions, this pressure is 1 atmosphere (atm) at sea level, where atm is a pressure measurement unit equal to 101325 Pascal, 101.325 kilo Pascal or 1.01325 bar. For diving applications, we frequently refer to the pressure at sea level as atmosphere absolute (ATA). It is not unusual to omit the fractions and assume that it's 1 bar.

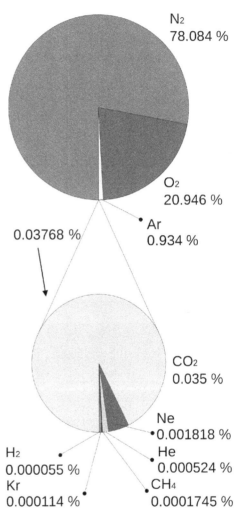

N_2
78.084 %

O_2
20.946 %

Ar
0.934 %

0.03768 %

CO_2
0.035 %

Ne
0.001818 %

H_2
0.000055 %

He
0.000524 %

Kr
0.000114 %

CH_4
0.0001745 %

Photo courtesy of Wikimedia Commons. Public Domain. "Atmosphere Gas Proportions."

The air pressure at the surface is approximately 1 bar, and we already know that air is composed of 21% O_2 and 79% N_2, we can presume that the *partial pressure of oxygen* at the surface (ppO$_2$) is 0.21 * 1 = 0.21 bar. Subsequently, ppN$_2$ would be 1 − 0.21 = 0.79 bar. To generalize, let's say that **the total pressure exerted by a mixture of gases (not in a state of reaction) is equal to the sum of the partial pressures of the individual gases**

composing this mixture. This is an established ideal gas law called *Dalton's Law* and not just an assumption. This law is a cornerstone in all diving applications.

When we dive, the water column above us exerts pressure on our bodies. The water's pressure increases by approximately 1 bar every 10 meters (33 feet). So at a depth of 10 meters (33 feet) the absolute pressure on our bodies is 1 bar (atmospheric pressure) + 1 bar (water pressure) = 2 bar. This pressure is also called the *ambient pressure*. Of course, the water's density varies from one place to another. For example, the Red Sea is denser than the Mediterranean, and salt water is denser than fresh water by approximately 2.5%. However, we'll neglect those fractions.

Now let's put this pressure-increase information along with Dalton's Law in perspective. At 10 meters (33 feet), the total pressure is 2 bar; $0.21 * 2 = 0.42$ of which is ppO_2, whereas the rest is ppN_2 $(2 - 0.42 = 1.58$ bar$)$.

The human body is composed of solids and liquids; both are incompressible. However, some natural and artificial cavities exist in the body. The natural ones are the sinuses, the middle ear, the lungs, the stomach, and the intestines; the artificial ones are mainly in the teeth (when bad fillings are involved). These cavities are filled with gas, and pressure affects gas volume. At constant temperature, when the pressure increases, the gas volume decreases, and vice versa. For example, at 10 meters (33 feet) the absolute pressure is doubled, and subsequently the volume is halved. This inversely proportional relationship is called *Boyle's Law* and is central in all diving applications.

Now let's compile all what we know so far in one table.

Depth		pp (bar)			
(m)	(ft)	O$_2$	N$_2$	P (bar)	V (relative)
Surface	Surface	0.21	0.79	1	1
10	33	0.42	1.58	2	1 / 2
20	66	0.63	2.37	3	1 / 3
30	100	0.84	3.16	4	1 / 4
40	130	1.05	3.95	5	1 / 5

Table 2.1: Partial pressures of oxygen and nitrogen in air at different depths, along with relative volume

Every system in the world is an effort to attain equilibrium, and the human body is no exception. At the surface, the body tissues are at equilibrium with the gases it does not metabolize. This means our tissues are saturated with nitrogen at the surface. When we start the dive (descend), the ppN_2 in the inspired gas increases with depth, so the body tissues try to attain equilibrium by absorbing the excess nitrogen. This process is called *ongassing*. On the ascent, at some point the process is reversed. The ppN_2 in the tissues becomes more than that in the inspired gas, so the tissues start releasing the excess nitrogen. This process is called *offgassing*.

In 1900, a Royal Navy diver descended to 45 meters (150 feet) in 40 minutes, spent 40 minutes at depth searching for a torpedo, then ascended to the surface in 20 minutes. Ten minutes later, he complained of abdominal pain and fainted. His breathing was labored, and he was cyanotic. He died after seven minutes. The next day an autopsy revealed healthy organs, yet the body had gas in the liver, spleen, heart, venous system, and subcutaneous fat.[1]

The body tissues and the blood could be represented as lipid (fat) and aqueous (water) substances. So the nitrogen they absorbed and released will obey the gases in solution law, also referred to as *Henry's Law*, which states that at constant temperature where no chemical reaction is taking place, the quantity of a gas that dissolves in a liquid is directly proportional to its partial pressure in the gas phase. This corresponds with the ongassing phase. Henry's Law also states that **when the partial pressure of the gas is reduced, a proportional amount of that gas will emerge from solution and may form bubbles in the liquid phase**. This corresponds with the offgassing phase. Bubble formation was the condition that Haldane, also known as "Father of Modern Decompression Theory," was trying to avoid at all cost. For decades, this was considered the main cause of DCS.

When nitrogen is breathed at elevated pressure, its partial pressure in the lungs initially exceeds that in the tissues. Nitrogen is then absorbed by the tissues until its partial pressure there equals that in the lungs. However, the tissues are not all the same. For example, the composition of the muscles is totally different than that of the fatty tissues. Also blood flows to various organs of the body in varying amounts. So we conclude that the principal factors governing the rate of nitrogen absorption by a particular tissue are perfusion (amount of blood flowed to the tissue) and solubility (of dissolved nitrogen in the tissue). Fatty tissues with high nitrogen solubility and poor perfusion gain and lose nitrogen more slowly than muscles with low nitrogen solubility and good perfusion. The process by which tissues uptake or eliminate nitrogen is called diffusion. This is how the dissolved nitrogen moves from an area of higher concentration to an area of lower concentration.

Halftimes

To handle the exchange, Haldane assumed the body was made of tissues with nitrogen uptake and elimination expressed as an exponential function. The tissue was half filled or half emptied in a constant time known as the *halftime*. The law of exponential decay, which is found throughout nature, was seen in the decay of radioactive material. For decompression modelers, perfusion, diffusion and solubility of a particular tissue could be expressed as a single parameter: the halftime of the tissue. Since the halftime tells us the rate at which the tissue uptakes or eliminates dissolved nitrogen, this is the all-in-one parameter we need for modeling the gas transport. No need to ask for specific details such as the solubility coefficients, perfusion, and/or diffusion rates.

Now let's consider the brain's white matter. This is a well-perfused tissue. Its halftime is about 5 minutes — i.e., it will become half filled with nitrogen in 5 minutes (50% *saturated*). In another 5 minutes, it will move halfway toward saturation, meaning that it will reach 75% saturation. Theoretically, the tissue will never be 100% saturated. Practically, in 6 cycles it will reach 98.4375% saturation, which is considered full saturation. So in 30 minutes, the brain's white matter would be considered fully saturated.

So if a diver made an immediate descent to 40 meters (130 feet) and stayed at this depth for 30 minutes, the 5-minute halftime tissue would be fully saturated. But how much nitrogen pressure will this tissue be holding? The air the diver breathes on the surface contains 0.79 bar ppN$_2$. Inside the lungs, the gas transport takes place through diffusion in the alveoli, which are the terminal ends of the respiratory tree. However, just as soap needs water to be effective, the alveoli need to be humidified by water vapor for that gas transport to take place. The result is that the *inspired ppN$_2$* is reduced to about 0.76 bar. This pressure is frequently called the *alveolar ppN$_2$*.

Back to our example, the dissolved nitrogen in the tissue is at equilibrium with alveolar ppN$_2$ on the surface. So the starting *tissue tension* is 0.76 bar. At 40 meters (130 feet), the absolute pressure is 5 bar, and the alveolar ppN$_2$ is 5 * 0.76 = 3.8 bar. The tissue is practically fully saturated after 6 cycles (30 minutes for a five-minute halftime tissue), so 3.8 bar would be the tissue tension after this period at that depth.

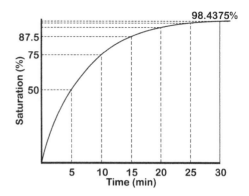

Time (min.)	Saturation
5	50%
10	75%
15	87.5%
20	93.75%
25	96.875%
30	98.4375%

Table 2.2: Tissue saturation can be expressed as an exponential function. Practically, full saturation takes place after 6 halftime periods.

Supersaturation, gradient, and critical supersaturation

When the ascent phase starts, the tension of the now fully saturated tissue will exceed the ambient pressure. This state is called *supersaturation*. To attain equilibrium, the tissue will tend to release the excess dissolved nitrogen in the venous blood, the venous blood will transport it to the lungs, and the lungs will discharge it out of the body through exhalation. The difference between the tissue tension and the ambient pressure is called the nitrogen *gradient*. The nitrogen release presumably will obey the same halftime concept of the intake.

To avoid caisson disease, the 1900 Royal Navy diver who died was using a very slow linear-ascent technique. He ascended from 45 meters (150 feet) to the surface in 20 minutes. Haldane realized this technique was flawed. Gas was found in the victim's organs, so he must have exceeded a *critical supersaturation* threshold. To allow gradual desaturation of tissues without bubble formation, Haldane presumed an absolute pressure reduction ratio of 2:1. Whenever this ratio is reached, the diver has to stop to release the excess nitrogen.[2] This technique resulted in progressively longer stops near the surface. These in-water decompression stops are traditionally at 3-meter (10-foot) intervals, with the shallowest stop being at 3 meters (10 feet). Now it's general practice to take the shallowest stop at 6 meters (20 feet).

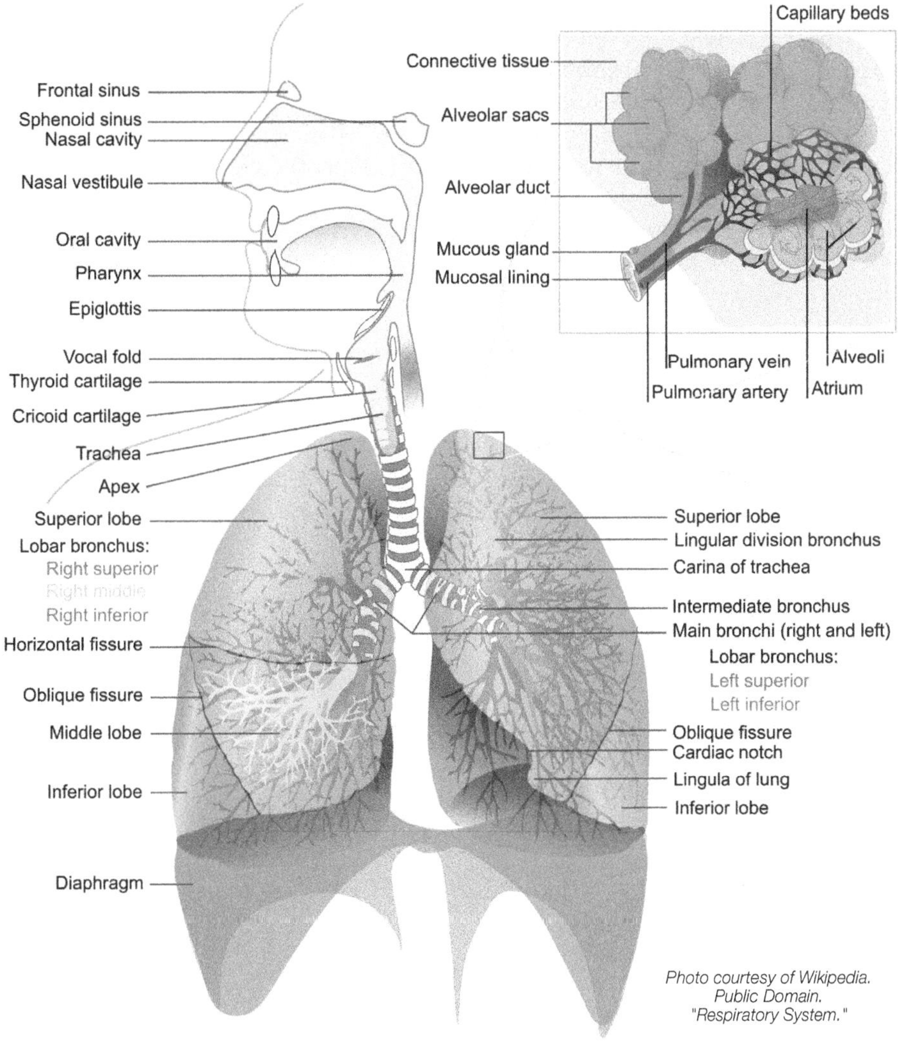
Capillary beds
Connective tissue
Frontal sinus
Sphenoid sinus
Nasal cavity
Alveolar sacs
Nasal vestibule
Alveolar duct
Oral cavity
Mucous gland
Pharynx
Mucosal lining
Epiglottis
Vocal fold
Thyroid cartilage
Cricoid cartilage
Pulmonary vein
Pulmonary artery
Alveoli
Atrium
Trachea
Apex
Superior lobe
Lobar bronchus:
Right superior
Right middle
Right inferior
Horizontal fissure
Oblique fissure
Middle lobe
Inferior lobe
Diaphragm
Superior lobe
Lingular division bronchus
Carina of trachea
Intermediate bronchus
Main bronchi (right and left)
Lobar bronchus:
Left superior
Left inferior
Oblique fissure
Cardiac notch
Lingula of lung
Inferior lobe

Photo courtesy of Wikipedia. Public Domain. "Respiratory System."

It's becoming clear that our "no-decompression" (or no-stop) dive is a special case, where the critical supersaturation threshold is not exceeded, thus no stops are required. The no-decompression limit (NDL) is the duration the diver can spend at a certain depth without incurring any decompression obligation because the supersaturation won't exceed the critical threshold. No-stop dives usually incorporate a "safety stop." This stop, which was introduced in the 1990s, interrupts the ascent to the surface with a 3- to 5-minute stage at 3 to 6 meters (10 to 20 feet). Doppler tests illustrated that a safety stop can reduce the incidence of venous gas emboli (VGE).[3]

Haldane also changed the ascent rates so that the first "jump" became faster. Modifications of Haldane's calculation methods are still the basis of many modern decompression schedules.

Ascent rates

Why did Haldane change the ascent rates if slow ascents were not at odds with decompression stops? He knew that the body was made of different tissues, so he presumed varying halftimes for different tissues. The longest tissue halftime Haldane proposed was 75 minutes. This would take 75 * 6 = 450 minutes to reach full saturation at a certain depth. So for a 30-minute, 40-meter (130-foot) dive, the 5-minute halftime tissue would be fully saturated, whereas the 75-minute one is not near saturation. With a slow ascent, the 75-minute halftime tissue would gain more nitrogen because its tension was still lower than the ambient pressure. So Haldane concluded that it's better to reach the first decompression stop as fast as possible.

Ascent rates are inherent parts of the decompression planning model. Normally there is an "accepted range" of ascent rates for each model. In 1993 the ascent rate to the first decompression stop was changed in the *U.S. Navy Diving Manual* from 18 meters/min. (60 feet/min.) to 9 meters/min. (30 feet/min.).[4] The effects of ascent rates can be complex; a rate too slow will allow more nitrogen to be absorbed at depth, whereas a rate too rapid will influence bubble formation. The range most modern decompression models use is 9 to 10 meters/min. (30 to 33 feet/min.).

Decompression researchers modified Haldane's calculations to cope with the evolving divers' requirements (such as extended immersion durations and greater depths). Now the "slow" tissue halftimes are frequently more than 600 minutes, which does not correspond with the body's physiology. So halftimes currently are just some mathematical parameters we use to predict DCS. The "tissue" halftime is a misnomer. We now call it *compartment* halftime. We'll discuss ultralong halftimes bridge in Chapter 8.

Summary

In this chapter, we discussed the basic principles of decompression. The "ingredients" of the air we breathe, the governing gas laws, the terms and definitions involved, along with the effects of pressure were detailed.

The concept of equilibrium as the motive for ongassing and offgassing came next. What diffusion, perfusion, and solubility are and how inert gas uptake and elimination by various body organs is thought to take place were portrayed. Subsequently, the concepts of halftimes, supersaturation gradients, and staging divers during the ascent phase were introduced.

CHAPTER 3

Dissolved-Gas (Haldanean) Models

Haldane

Based on his studies on goats, Haldane came up with three basic assumptions. First, the progress of saturation with nitrogen in mammals breathing compressed air generally follows the line of an exponential curve. Second, the curve of desaturation after decompression is symmetrical to that of saturation. Third, the time in which a mammal exposed to compression becomes saturated with nitrogen varies in different parts of the body.

The first two assumptions were modeled by what we know now as the *Haldane equation*. This is the formula we have used (until now) to calculate the inert gas uptake or elimination at any constant depth.

The third assumption was modeled by introducing what we now call compartment halftimes. Haldane presumed the human body would reach full nitrogen saturation in five hours. Different rates were assigned to hypothetical "tissues." This supposedly simulates the variation in nitrogen uptake and elimination rates in different body organs. Haldane suggested 5 independent hypothetical tissues with halftimes of 5, 10, 20, 40, and 75 minutes. The 75-minute halftime would reach 93.75% of full saturation in 5 hours.

Haldane observed that a diver breathing compressed air can ascend from a 10-meter (33-foot) depth (2 ATA) directly to the surface (1 ATA) without getting bent. He deduced that a pressure reduction ratio of 2:1 would be safe, and he used that ratio to stage divers during ascent. The initial phase of the ascent was characterized by a relatively rapid rate from depth to the depth of the first decompression stop (which he assumed to be half the absolute pressure on the diver), then followed by a noticeably slower ascent rate to the surface. British physiologist Sir Leonard Hill (1866–1952) strongly disagreed with Haldane's approach and theorized decompression should use a slow, linear ascent to the surface. Using goats, Haldane was able to prove a slow, linear ascent was actually unsafe because too much nitrogen remained upon surfacing, which in turn resulted in frequent DCS hits.[1]

It is worth noting that Haldane's model is *perfusion limited*, meaning it assumes that the hypothetical tissues are independent to each other yet are connected *in parallel* to the blood flow. The model presumes the blood comes from the heart, is divided by the arteries, passes through the tissues, and finally enters the veins to return to the heart. No gas transfer from one tissue to another is considered, so different tissues do not affect each other, neither by diffusion nor by any other means. Haldane presumed all tissues have the same characteristics, except for the varying amounts of blood flow (perfusion), which led to the different rates of nitrogen exchange (halftimes). This isolated arrangement assumes that all the hypothetical tissues are risk bearing, meaning that DCS could occur at any one of them. According to these assumptions, Haldane developed the first decompression tables.

Workman

Workman suggested Haldane's original pressure reduction ratio of 2:1 was a ratio of 1.58:1 because only the partial pressure of nitrogen should be considered. In other words, at 10 meters (33 feet), the partial pressure of nitrogen (ppN_2) is 0.79 * 2 = 1.58 ATA, not 2 ATA. If a diver on air (79% nitrogen) can stay indefinitely at a 10-meter (33-foot) depth and not get bent upon surfacing (this assumption later was found to be inaccurate), will this hold true if, for example, the oxygen in the breathing mix was reduced from 21% to 16%?

The revised U.S. Navy Standard Air Decompression Tables came into use in 1956. To expand their range and make them suitable for deeper dives, the model used to "cut" these tables advocated the use of six compartments rather than five. The model employed the first four compartments originally proposed by Haldane along with a slightly altered fifth compartment (80-minute halftime instead of 75) and a much slower sixth compartment (120-minute halftime). In 1957 Workman illustrated that this pressure reduction ratio not only varied from one halftime compartment to another, but it also varied with depth. He added three compartments with 160-, 200- and 240-minute halftimes. With 1,507 total man-hours of test dives, analysis of the data he collected confirmed that the "faster" compartments tolerated greater overpressure ratios than the "slower" ones, a conclusion Hawkins, Shilling, and Hansen made 22 years earlier. Workman's analysis also highlighted that the tolerated overpressure ratios decreased with increasing depth for all compartments.[2]

In 1965 Workman extended his work to include helium. Instead of using ratios, he described the maximum tolerated partial pressure of nitrogen and helium for each compartment at any given depth as well as at the surface and coined the term "M-values" for them, where M stands for maximum. In the form of a linear equation, he presented these M-values as a function of depth (ambient pressure).[3] He argued this "linear projection" was useful for computer programming. (In the 1960s computing power and memory were serious issues.)

Bühlmann

Instead of trying to reinvent the wheel or attempting to come up with new concepts, Bühlmann compiled the reliable work of other researchers and invested his time in perfecting the model to minimize the probability of DCS occurrence. Bühlmann adopted Haldane's assumptions and used Haldane's equation to calculate the inert gas uptake or elimination at any constant depth. He communicated frequently with Dr. Heinz Schreiner (1930–1996) of the Linde Research Laboratory. Working with programmer Patrick Kelley, Schreiner built on the experience of the British and the U.S. navies and significantly contributed to decompression modeling by solving the differential equation for gas exchange when the ambient pressure changed at a constant rate. Bühlmann incorporated a subset of this equation to calculate the inert gas uptake or elimination at any variable depth (ascent and descent segments). Schreiner also established an important concept: The total inert gas pressure (total load) in a given compartment is the sum of the partial pressures of all inert gases in that compartment, even if they have different halftimes. We'll see how useful that concept is when we discuss mixed gases in Chapter 5.

Bühlmann was Swiss, so his calculations always considered diving at altitude (in the Swiss high mountain lakes). He knew that the inspired gas was diluted by water vapor and carbon dioxide (CO_2) in the lungs and that their partial pressures, which could be neglected for sea-level diving, would become significant for diving at altitude. To account for that, the *alveolar inert gas pressure* should be considered rather than the inspired inert gas pressure. In 1971 Schreiner proposed a respiratory quotient of 0.8.[4] This would lead to an approximate 0.0493 bar reduction in the ambient pressure. Bühlmann, however, used a respiratory quotient of 1, which would lead to an approximate 0.0627 bar reduction in the ambient pressure. The U.S. Navy used a respiratory quotient of 0.9, which would lead to an approximate 0.0567 bar reduction in the ambient pressure. Schreiner's value was the most conservative, whereas Bühlmann's value was the most aggressive. Ultimate Planner uses Schreiner's value for decompression calculations only. To represent the worst-case scenario, most decompression programs do not subtract the water vapor pressure from the ambient pressure when calculating the ppO_2 for estimating the total CNS.

Relying heavily on Workman's work, Bühlmann conducted research over a span of some 35 years to adjust the M-values for safer decompression. To suit a wider variety of diving applications (including diving at altitude), he based his M-values on absolute pressure rather than depth pressure. He also modified Workman's linear equation and expressed it in the form of (a) and (b) coefficients. However, it's easy to convert back and forth between Bühlmann's newer style and Workman's traditional style.

Bühlmann based his M-values on absolute pressure rather than depth pressure. This did not mean his models were viable for space (0-g). Although the Bühlmann M-value line started at zero absolute pressure, this was only for mathematical representation. Just like Workman M-values, Bühlmann M-values are for scuba divers, not astronauts. The M-values are capable of computing the decompression requirements for diving at both sea level and altitude. The lowest supersaturation at which DCS can occur, independent to diving, is about 0.5 atm, which corresponds to an altitude of about 5,486 meters (18,000 feet).

Workman style:

M = delta M * D + M(0)

D is the depth.

Workman to Bühlmann style:

Coefficient (a) = M(0) – delta M * Ambient Pressure (absolute)

Coefficient (b) = 1 / delta M

Bühlmann increased the number of compartments to 16. He proposed the ZH-L12 model (ZH stands for Zürich, L for limits or linear, and 12 for 12 pairs of coefficients). Although this model

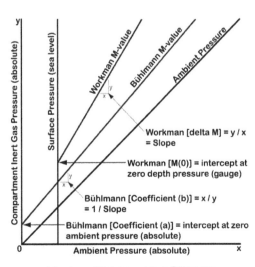

Workman [delta M] = y / x = Slope

Workman [M(0)] = intercept at zero depth pressure (gauge)

Bühlmann [Coefficient (b)] = x / y = 1 / Slope

Bühlmann [Coefficient (a)] = intercept at zero ambient pressure (absolute)

M-values: Workman versus Bühlmann

uses 16 pairs of compartment halftimes, it uses only 12 pairs of coefficients. The 10th and 11th compartments have similar M-values, as do the last 4 compartments. ZH-L12 was published in Bühlmann's 1983 book *Dekompression: Dekompressionskrankheit* (in English, *Decompression: Decompression Sickness*). Years later he proposed the ZH-L16 model, which is quite popular as a basis for decompression planning. ZH-L16 was first published in his 1990 book *Tauchmedizin*, or *Diving Medicine*.

Both ZH-L12 and ZH-L16 cover a range of halftimes up to 635 minutes. A tissue with a blood flow so limited to create such a long a halftime would be oxygen deprived. (These ultra-long halftimes are discussed more in Chapter 8). It's now clear that both models are entirely arbitrary in the sense that they represent the actual physical processes in the diver's body. Instead, they are mathematical attempts at modeling the empirical results that Bühlmann collected over the years.

The original ZH-L16 algorithm (frequently referred to as ZH-L16A) was found to be a bit aggressive in the middle compartments, so Bühlmann proposed a model variation called ZH-L16B to make it slightly more conservative. This variation is recommended for planning dives (cutting tables). He also proposed another considerably more conservative variation called ZH-L16C, which is more convenient for real-time usage in dive computers.

One of the key concepts Bühlmann demonstrated is the tolerated partial pressures of two different gases in the same compartment will vary according to their solubility coefficients in the *transport medium* that delivered those gases to that compartment. The transport medium in our case is the blood plasma.[5] Using this concept along with *Graham's Law* (which states the volume of gas diffusing into a liquid is inversely proportional to the square root of the molecular weight of the gas), complete sets of halftimes and M-values for other gases (such as helium) could be derived mathematically. Bühlmann also stated the overall M-value for a given compartment with multiple gases, each gas having a different M-value, will vary according to the proportion of each gas present in the compartment, which means the computations will be in accordance to the partial pressures of the different inert gases. These concepts are discussed further in Chapter 5.

Four editions of Bühlmann's book were published from 1983 to 1995. They were considered the only nearly complete references on making decompression calculations, and the model specifics were released in the public domain. That's why ZH-L16 became the basis for most of the market's decompression-planning software tools and dive computers. In turn, divers have performed millions of dives using that algorithm and have learned how to use it considerably safely.

Adaptive algorithms
In the 1990s Bühlmann wanted to reflect the change in blood perfusion to various body organs when the diver is subjected to temperature and/or workload variations, as changes in blood perfusion would alter the inert gas saturation tolerance — i.e., he wanted to develop an algorithm to deal with all the real-time variables throughout the dive, not just the depth and time. The result was ZH-L8 ADT, a model using 8 evenly distributed tissue compartments with halftimes ranging from 5 to 640 minutes (for nitrogen). Some of the initial halftimes (particularly those of the mid compartments) get altered

during the dive to reflect what's going on with the diver's body. The idea is that coldness causes vasoconstriction (mainly at the skin and the muscles), thus reducing perfusion. To mathematically simulate this change in perfusion level, the corresponding halftimes should be altered.

This model is useful only for dive computers. The dive computer has a sensor to monitor the water temperature throughout the dive, then the feedback is sent to the algorithm. One drawback of this approach is the water temperature is not always a measure of what the diver is feeling. The diver might be wearing a swimsuit and a T-shirt or a drysuit with a heavy undergarment, so skin temperature would have been a more appropriate indicator. The dive computer calculates the workload by monitoring the reduction in tank pressure (gas-consumption rate), which some divers consider another drawback because an increased consumption rate is not always an indication of elevated workload. The newer, top-line models include an optional integrated heart-rate monitor — a belt that allows the workload calculation to take into account actual blood circulation. Some models have the workload as an adjustable setting, which is good for unfit divers or when the diver knows beforehand that there will be a demanding situation (swimming against current, a long surface swim before the descent, etc.).

Based on his personal communication with Bühlmann and Max Hahn, my friend Dr. Albrecht Salm gave me some details about which ZH-L16 halftimes would need to be altered to reflect cold and/or an increased workload. Ultimate Planner introduced ZH-L16D, a more conservative model than both ZH-L16B and ZH-L16C that generates more suitable schedules for unfit divers or for anticipated colder and/or more demanding dives. For example, assuming the last stop depth is at 6 meters (20 feet), the total run time of a 30-minute dive on air to a 45-meter (150-foot) depth would be 73 minutes (ZH-L16B), 84 minutes (ZH-L16C), or 87 minutes (ZH-L16D).

Deep stops

During the 1980s and the early 1990s, Dr. Richard Pyle performed scores of dives in the 55- to 67-meter (180- to 220-foot) range. The number of dives was statistically significant for him to notice certain patterns, one being he would frequently feel tired, unwell, or uneasy to some extent following deeper dives. Since it had nothing to do with thermal protection or the level of physical exertion, it was obvious to him the symptoms of this postdive fatigue were associated with inert gas loading. He discovered a subpattern showing the fatigue following deeper dives was inconsistent – it didn't occur after all dives. In an attempt to correlate the severity of symptoms with possible DCS contributing factors such as dehydration, exercise before diving, water temperature, etc., he managed to get a match: Pyle, an ichthyologist, didn't experience any postdive fatigue on dives when he collected fish.

By examining the dives on which he collected fish, Pyle ascertained the level of exertion on these dives would normally be higher than comparable profiles when he didn't catch anything, because chasing fish would require more work at depth. However, he made a more significant observation. To manage their buoyancy in the water column, most fish species have a gas-filled internal organ (swim bladder). To vent off the excess gas, Pyle needed to stop at some point during the ascent and temporarily insert a hypodermic syringe into the fishes' swim bladders to prevent them from expansion, thus avoiding damage to the adjacent organs. Typically, he would need to stop for two or three minutes at a much deeper level than the first obligatory stop dictated by his decompression-planning tool.

Pyle's observations were against the conventional wisdom of the dissolved-gas models in use at the time; a couple of minutes at a greater depth than that of the first stop would only cause more decompression stress.

Pyle started employing his "deep stops" strategy on all the dives that required decompression stops. He later reported all symptoms of fatigue virtually disappeared. Subsequently, he started communicating his findings to the dive community. His approach was met with considerable skepticism, and he was sometimes lectured by some self-appointed guardians of the traditional decompression theory.

At an American Academy of Underwater Sciences (AAUS) meeting in 1989, Pyle saw a presentation by Yount about his VPM (which will be discussed in Chapter 6), and it finally started to make sense. With the increasing use of Doppler tests in decompression-related research around the world, consistent results showed that bubbles are present in the body after almost all dives, leading to the emergence of the asymptomatic (silent) bubbles concept. Pyle then was able to articulate that his deep stops would help control the size of whatever bubbles were formed in the body during the initial phase of the ascent. He related his deep stops to the safety stops divers frequently take on no-stop dives, asserting that while those nonrequired stops are usually at a 6-meter (20-foot) depth, they're actually considered deep stops since the dive required none.

Pyle developed his own protocol to calculate deep stops (or deep safety stops) for any dive profile. Frequently referred to as Pyle stops, the protocol is as follows:[6]

1. Calculate the decompression schedule for the dive using your preferred software tool.

2. Find the midpoint between the bottom portion of the dive (at the time you begin your ascent) and the first required decompression stop.

3. Stay at this depth for two to three minutes.

4. Recalculate the decompression schedule after inserting the deep stop in the profile as a depth segment.

5. If the distance between your deep stop and the first required decompression stop dictated by the recalculated schedule is greater than 9 meters (30 feet), add another deep stop at the midpoint between the previous deep stop and the first required stop.

6. Repeat as necessary until there is 9 meters (30 feet) or less between your last deep stop and the first required safety stop.

For example, using Bühlmann's ZH-L16B to plan a 30-minute dive to a maximum depth of 55 meters (180 feet) results in a first required decompression stop at 15 meters (50 feet). The midpoint is at (55 + 15) / 2 = 35 meters (115 feet), which is where the first Pyle stop should be inserted. Now add to the dive profile 2 or 3 minutes at 35 meters (115 feet) and regenerate the dive schedule. Let's assume the first deco stop is still at 15 meters

(50 feet). The difference between the first deco stop and the inserted deep stop is 35 − 15 = 20 meters (66 feet), which is greater than 9 meters (30 feet). So repeat the former step, and insert the second Pyle stop at (35 + 15) / 2 = 25 meters (82 feet). By adding 2 or 3 minutes at 25 meters (82 feet) to the dive profile and regenerating the dive schedule, let's assume the first deco stop is shifted down 3 meters (10 feet), so it's now at 18 meters (60 feet). Now the difference is 25 − 18 = 7 meters (23 feet), which is less than 9 meters (30 feet), so we can exit the loop.

Another way to compute where to insert arbitrary deep stops is to employ the gas volume expansion (GVE) technique, which assumes the gas volume in your body will decompress during the descent and the bottom time and you should stage your ascent according to an arbitrary percentage. With experience, you can fine-tune this percentage according to your preference and the dive conditions. As with Pyle's approach, the stop duration is 2 or 3 minutes, and the process is repeated until the difference between the last inserted stop and the first required deco stop becomes 9 meters (30 feet) or less.

Back to our 55-meter (180-foot) dive where the first required deco stop is at 15 meters (50 feet), assuming we'll use GVE = 40%, we divide the absolute pressure by the initial GVE at that depth (always equal to 1) plus the arbitrary staging percent (0.4). So the first GVE stop would be at 6.5 / (1 + 0.4) = 4.64 bar = 36 meters (118 feet). Now add to the dive profile 2 or 3 minutes at 36 meters (118 feet) and regenerate the dive schedule. Assuming the first deco stop is still at 15 meters (50 feet), the difference between the first deco stop and the inserted deep stop is 36 − 15 = 21 meters (70 feet), which is greater than 9 meters (30 feet). So repeat the former step and insert the second GVE stop at 4.6 / 1.4 = 3.28 bar = 23 meters (75 feet). Add 2 or 3 minutes at 23 meters (75 feet) to the dive profile and regenerate the dive schedule. Assuming that the first deco stop is shifted down 3 meters (10 feet), it's now at 18 meters (60 feet). Now the difference is 23 − 18 = 5 meters (16 feet), which is less than 9 meters (30 feet), thus we're done.

The arbitrary values divers use for GVE normally range from 20% to 70%. Neither protocol considers the *start of deco zone* (the depth at which at least one compartment begins to offgas). For example, if we applied GVE = 25% on our example dive, we'll get the first stop inserted at 6.5 / 1.25 = 5.2 bar = 42 meters (138 feet). The start of deco zone for this dive is at 40.6 meters (133 feet), which means all compartments will be ongassing at 42 meters (138 feet) — disappointing divers who believe the main advantage of inserting arbitrary deep stops is to reduce the tension in the fast tissues to stop bubbles from forming in the first place (rather than control the size of the already present bubbles).

In 2004 in an attempt to study the effect of introducing a deep stop, 22 volunteers made a total of 181 dives. The study concluded that the introduction of a deep stop significantly reduced Doppler-detected bubbles, together with tissue gas tensions in the 5- and 10-minute halftime compartments.[7]

Gradient factors
Bühlmann's last edition of his book *Tauchmedizin* (*Diving Medicine*, but available only in German) was published in 1995. Some DCS incidents occurred even though the victims had surfaced with less tissue loadings than the proposed M-values (some at around 90% of the M-values), indicating the M-values Bühlmann proposed were in

some cases very aggressive. The situation became worse for repetitive dives. Bühlmann acknowledged that and cautioned that reduction factors must be applied for repetitive diving calculations. In general, applying some degree of conservatism appeared prudent.

The software developers who incorporated Bühlmann's algorithm in their programs at first used some fudge factors to generate more conservative profiles, and it worked to some extent (see Chapter 8). But coupled with the anecdotal data that favored deep stops and the techniques the diving community used to generate them (mainly Pyle and GVE), planning dives with ZH-L algorithms became hectic and to some extent unreliable. The diver needed to insert some arbitrary deep stops, along with applying "something" to generate more conservative schedules. The need for a consistent *M-value reduction mechanism* across the entire ambient pressure range became obvious.

The diver starts the ascent, obviously from the ambient pressure line, until reaching the M-value line. To not go past the M-value line, the diver has to stop for some time (deco stop) until the tissue tension is reduced to the extent the diver can ascend several meters or feet (deco step size) without passing the M-value line. This process continues until the diver reaches the surface.

To both introduce a deeper first stop and apply some degree of conservatism, we can shift the M-value line toward the ambient pressure line to *trim the decompression zone.* Professional engineer Erik Baker introduced this approach, which he called *gradient factors* (GF), in the late 1990s. For example, trimming the decompression zone by 30% moves the M-value line toward the ambient pressure line so that the resulting area equals 70% of the original decompression zone — a 70% GF. The initial jump from the ambient pressure line to the newer GF line is shorter than that to the original M-value line, meaning the initial pull in the water will be also shorter, introducing a deeper first stop. The time of the shallower stops would be extended because the diver now ascends with 70% of the allowable tissue tensions.

This approach is a two-edged sword. To make the introduced stops deep enough, you would lower the GF, which will greatly extend the shallower stops. So Baker separated the GF into two lines: GF low and GF high. GF low is closer to the ambient pressure line and is responsible for introducing the deep stops, whereas GF high is closer to the M-value line and is responsible for applying conservatism. This technique greatly smooths the ascent profile. GF 100/100 takes you back to the raw model. Divers choose the GF combination they feel more comfortable with and

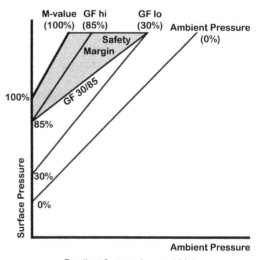

Gradient factors: low and high

frequently alter their preference according to the dive conditions. For example, some divers don't like starting their stops very deep in cold water, so they might increase the GF low for cold-water dives and decrease it when diving in tropical waters. Other divers tend to increase the GF high above the 100% bar, which would lead to very aggressive profiles not prescribed by the decompression model. These divers claim that introducing deeper stops would suppress bubble formation and expansion so no need to extend the shallower stops; actually trimming them might be more prudent. We'll examine this hypothesis in detail in Chapters 6 and 8. A popular combination among divers is GF 30/85.

Although gradient factors are a mathematical approach to trimming the decompression zone, they are still considered an arbitrary approach.

Summary

In this chapter, we discussed how Haldane managed to develop the first decompression tables, how Workman extended Haldane's work and introduced the M-values concept, and how Bühlmann "compiled" the reliable work of other researchers to perfect his algorithms. We also discussed Bühlmann's ZH-L12 model, his most popular ZH-L16 model and its variants, along with his adaptive ZH-L8. Then we explained the origins of the deep stops approach and a couple of methods to compute where to insert arbitrary deep stops. Finally, Baker's gradient factors were illustrated as a consistent M-value reduction mechanism across the entire ambient pressure range, introducing deep stops and adding conservatism.

CHAPTER 4
Nitrox

The introduction of nitrox to the sports-diving community in the early 1990s was not as easy and seamless as one might have expected. Bret Gilliam explains the controversy as follows:

"Suddenly, we turned the corner on the 1990s, and a peculiar mood began to intrude on diving. I still look back now some 20 years later and am amazed at some of the nonsensical attitudes that tried to squelch anything innovative or new that came forward from the self-appointed 'guardians' of diving's old school. If you weren't around then to remember the incredible controversies that abounded over anything associated with technical diving, nitrox, or even how many dives a day was considered allowable, let me take you in my little 'virtual time machine' and revisit the era when a segment of arch-conservatives tried to hijack diving's future . . .

"Dick Rutkowski was the first target. When he retired as NOAA's (National Oceanic and Atmospheric Administration's) Deputy Director of Diving, he decided to expand his cutting-edge training programs on recompression treatments and chamber operations to include the first formalized training in nitrox. Nitrox had been around for two decades but was largely limited to commercial and science diving applications. But Rutkowski saw the huge benefit for sport divers to lengthen no-decompression windows and allow more dives per day by shortening surface intervals. The advent of the first reliable dive computers also signaled a future for divers to program selected O_2 mixes for optimum dive efficiency.

"But you would have thought that he suddenly came out in favor of child molestation in the response he received from the conservatives. DAN's (Divers Alert Network's) executive director Peter Bennett condemned him as an irresponsible reprobate who should be thrown out of diving professionally. Skin Diver magazine's editor Bill Gleason ran a sensational editorial referring to nitrox as 'The Devil's Gas.' PADI and SSI proclaimed that nitrox had no place in recreational diving. I was on NAUI's board of directors at the time and suggested that NAUI lead the way and get behind the new technologies such as nitrox, dive computers and technical diving in general. My perspective fell on deaf ears, so I joined with Tom Mount and Billy Deans to expand Rutkowski's International Association of Nitrox Divers (IAND) and widen our curricula as the International Association of Nitrox and Technical Diving (IANTD). We were immediately branded as the equivalent of heretical witches and targeted as leading the whole industry on a path to destruction.

"In a way, it was kind of amusing. People who had already 'been there, done that' like Giddings, Waterman, Exley, Hollis, the Taylors, would occasionally check in to see what bombs were being lobbed in our direction, but all simply advocated getting the information out there for the public to make an informed choice. Mount, Rutkowski and I became the designated spokespersons for much of the technical community in endless rounds of dive show presentations and workshops. At times even the opportunity for an equal chance to reach the public was denied. The 1991 Diving

Equipment and Manufacturing Association (DEMA) board briefly voted to ban all exhibits for nitrox and technical diving training and products. This was swiftly withdrawn after we had a legal dialogue on things like 'restraint of trade' and 'tortious interference in business.' So all was reinstated in time for the 1991 trade show.

"In early 1994 I departed IANTD and formed Technical Diving International (TDI). I was also by then the chairman of the board of NAUI and reached out to other prominent and credible diving professionals to bring the message of the newer technologies to as wide an audience as possible. But it still seemed incredible to us that anyone with an IQ above room temperature could not see the benefits of nitrox, diving computers, modern rebreathers and the establishment of proper training programs to ensure that divers had the chance to get educated properly. In late 1995, I took over Uwatec as vice president and CEO and began yet another parallel business expansion in manufacturing diving instruments, computers and rebreathers.

"Finally, the real reasons for all the bomb-throwing began to become clear. It was about the conservatives' mistaken opinions about their supposed influence and ability to control the industry. Boiled down to its essence, it was about power and magazine ad sales.

"Because it seems that the diving public was a lot smarter than Skin Diver and their henchmen understood. When accurate articles with proper information were published, the divers voted with their wallets and decided to take nitrox and technical training, buy diving computers and newly designed gear like streamlined back-mounted BCDs, enlarged volume cylinders and other products that just made sense to a diver's ability to maximize their enjoyment of the sport and their safety.

"The whole house of cards tumbled down in 1996 when PADI finally caved in and started nitrox training; they were the last of the major agencies to do so.

"The entire era of extreme controversy barely lasted eight years. All of the naysayers were proven either to be totally uninformed on the subject intellectually or to have their own hidden personal agendas. Nitrox, diving computers and technical diving became mainstream and widely accepted everyday practice worldwide.

"As Dick Rutkowski famously said, 'Science always triumphs over bullshit.' He was right. But the battles made things interesting for a while." [1]

This was taking place on the western side of the Atlantic. On the eastern side, Kevin Gurr, Richard Bull, and the late Rob Palmer formed the European Association of Technical Divers (EATD) and offered the first nitrox courses in Europe in 1992. The Sub-Aqua Association (SAA) recognized nitrox in 1994, and the British Sub-Aqua Club (BSAC) followed one year later.

Now let's take a close look at "the devil's gas."

We know that air at sea level contains almost 21% oxygen. The other 79% is mainly nitrogen (78% +), along with approximately 1% of other gases (primarily argon). For simplicity, let's discard the other gases and consider it an oxygen–nitrogen mixture. Nitrox is theoretically any oxygen–nitrogen mix and practically any oxygen–nitrogen mix where oxygen is more than 21%. Nitrox is also called oxygen enriched air (OEA) and enriched air nitrox (EAN). EAN is the most commonly used term and is abbreviated as EANx, where x indicates the percentage of oxygen in the mixture. For example, EAN36 is 36% O_2 and 64% N_2.

The oxygen percentage in any nitrox mix is practically more than that in air, so nitrox is not a deep-diving gas. The possibility of oxygen toxicity increases because of these elevated levels of oxygen. That's why nitrox has depth limitations and should be used at shallower depths than air. One unfounded myth associated with "the devil's gas" is that it affects recompression treatment because the patient would be at risk of pulmonary oxygen toxicity. To counter this myth, oxygen loadings have been tracked correctly. All maximum daily allowances were found to permit a complete therapeutic treatment whenever needed.

When compared with air, nitrox has the advantage of lowering the tissue loadings after diving and subsequently shortening the surface intervals between dives. Divers breathing nitrox frequently report less postdive fatigue. Less nitrogen content means extending the no-stop time (also known as no decompression limit, NDL) and reducing the decompression obligation if any. That's why nitrox use is a significant element in the reduction of DCS risk. Using nitrox also means reducing the undesirable nitrogen narcosis effects.

Nitrox physiology

Because they are composed of the same gases, nitrox and air have the same physiological effects on divers. However, nitrox contains more oxygen and less nitrogen, so the effects occur at different depths and at uneven time intervals. As the nitrox diver descends, the higher oxygen content results in a higher partial pressure of oxygen (ppO_2) and the lower nitrogen content results in a lower partial pressure of nitrogen (ppN_2) than diving on air. The ambient (surrounding) pressure at any given depth is the same in both cases.

Depth		ppO_2 (bar)		ppN_2 (bar)		P (bar)
(m)	(ft)	Air	EAN40	Air	EAN40	
Surface	Surface	0.21	0.4	0.79	0.6	1
10	33	0.42	0.8	1.58	1.2	2
20	66	0.63	1.2	2.37	1.8	3
30	100	0.84	1.6	3.16	2.4	4
40	130	1.05	2	3.95	3	5

Table 4.1: Partial pressures of oxygen and nitrogen in air and EAN40

Henry's Law states that the amount of gas dissolved in a liquid is directly proportional to the partial pressure of the gas in contact with the liquid. By applying this law we can easily conclude that for the same dive profile, breathing nitrox rather than air means less nitrogen will dissolve in the blood. Subsequently, less nitrogen is carried to the body tissues, which means a lower level of ongassing.

To get a better picture, let's consider a dive to a depth of 25 meters (83 feet) on EAN40. The ppN_2 is simply the ppN_2 on the surface (0.6 bar) multiplied by the ambient pressure at 25 meters (83 feet), which is 3.5 bar. The result is 0.6 * 3.5 = 2.1 bar.

Now let's determine the depth at which ppN_2 = 2.1 bar if the diver were breathing air. To do that we have to:

1. Divide the ppN_2 at depth by that on the surface (0.79 bar): 2.1 / 0.79 = 2.66 bar.

2. Convert this ambient pressure to depth: 2.66 bar = 16.6 meters (55 feet).

This means breathing EAN40 at 25 meters (83 feet) subjects the diver to the same partial pressure of nitrogen, and accordingly the same decompression obligation, as breathing air at 16.6 meters (55 feet). In other words, the equivalent air depth (EAD) of EAN40 at 25 meters (83 feet) is 16.6 meters (55 feet).

We can calculate the EAD for any nitrox mix at a given depth by using this general formula:

$$\textbf{EAD (m)} = ([(1 - \textbf{fO}_2) / 0.79] * [\textbf{Depth in meters} + 10] - 10), \text{ or}$$

$$\textbf{EAD (ft)} = ([(1 - \textbf{fO}_2) / 0.79] * [\textbf{Depth in feet} + 33] - 33)$$

If you want to dive nitrox yet prefer to plan your dives using air tables, calculate the EAD and plan accordingly. This will ensure that at the end of your bottom time you have ongassed the same amount of nitrogen as if you were diving on air to the shallower EAD. However, on ascending you're still breathing less ppN_2 than that of air, which means that you're offgassed faster. This adds a safety margin. If you want to add yet another safety margin, simply breathe nitrox but plan your dive on air tables using the actual depth (not the EAD). Similarly, you can breathe nitrox yet leave your dive computer settings adjusted to air.

A study was performed in 2010 in the resort town of Sharm El Sheikh, Red Sea, Egypt, to evaluate the perceived fatigue in divers after air and nitrox diving.[2] The research studied fatigue in 219 healthy divers performing either an air or an EAN32 dive to 21.2 ± 4 meters (70 ± 13 feet) for 43.3 ± 8.6 minutes. The two groups were comparable in sex ratio, age, and diving experience. Divers were assessed predive and 30 to 60 minutes after surfacing using a visual analog scale (VAS) of fatigue and critical flicker fusion frequency (CFFF). The change in perceived fatigue level after a single dive was significantly lower when divers breathed EAN32 compared with air. The study concluded three hypotheses were to be considered to explain the difference in postdive fatigue: an oxygen effect, a nitrogen effect, and a bubble effect. The study stated these hypotheses involve complex phenomena in the functional modifications of the nervous system at elevated pressures, thus more research was required to explain them.

We've already looked at the nitrogen effect, so let's look at the effect of oxygen and bubbles. In 2012 a study was performed to evaluate the effects of successive air and nitrox dives on human vascular function.[3] The aim of this study was to evaluate the changes in vascular functions following scuba diving and to assess the potential differences between two breathing gases: air and EAN36. Ten divers performed two 3-day diving series (a no-stop dive to 18 meters (60 feet) with a 47-minute bottom time using air and nitrox, respectively) with two weeks pause in between. Ultrasonic determination of venous gas bubble grade assessed production of nitrogen bubbles postdive. The study concluded that significantly higher bubbling was found after all air dives as compared with nitrox dives. The study also concluded that nitrox diving affects the vascular functions more profoundly than air diving by reducing flow-mediated dilation (FMD) response, probably due to higher oxygen load.

The bottom line is that the effects of using nitrox may not be solely due to the decreased amount of the inhaled nitrogen. Increased oxygen content is established to have vascular effects, and the amount of bubbling is significantly affected. More studies are required to accurately determine the effects of higher oxygen content and decreased bubbling.

One quick note about oxygen: To support human life indefinitely, a breathing mixture must have ppO_2 between 0.16 and 0.5 bar. Less than 0.16 will cause hypoxia (oxygen starvation), and more than 0.5 eventually leads to oxygen toxicity.

Taking a look back at our 25-meter (83-foot) dive on EAN40, the ppO_2 is 1.4 bar — obviously higher than the 0.5-bar mark. Unless the ppO_2 and exposure time are kept within certain limits, oxygen toxicity will occur. Oxygen toxicity manifests itself in two forms: acute CNS oxygen toxicity and chronic pulmonary oxygen toxicity. Chronic pulmonary oxygen toxicity (sometimes referred to as "Lorrain Smith effect") is a long-term effect of breathing rich oxygen concentrations for extended periods, so in our case it's not considered much of a problem. On the other hand, the effects of acute CNS oxygen toxicity are hazardous to the diver. As the ppO_2 increases, oxygen solubility in the serum increases, so the toxicity of oxygen increases. If the oxygen exposures are not tracked accurately, this increase in toxicity could lead to convulsing underwater and subsequent drowning and death.

Nitrox for accelerated decompression

Consider a 30-minute dive on air to a maximum depth of 45 meters (150 feet). Using air only, the total run time is 104 minutes, assuming ZH-L16B with gradient factors 30/85 and the last stop at 6 meters (20 feet). But if you use EAN60 starting at 15 meters (50 feet), the total run time becomes only 63 minutes.

Ultimate Planner 1.5 by Asser Salama.

Warning: This software is intended for demonstration purposes only. The author accepts absolutely no responsibility for the schedules generated by this software. Use it at your own risk.

Buhlmann-GF/U: OFF
Model: ZH-L16B
GF Low: 30% - GF High: 85%
Altitude: 0.0m
Leading compartment enters the decompression zone at 33.0m
Run time includes the ascent time required to reach the stop depth

Depth	Seg. Time	Run Time	Mix	ppO2	Depth	Seg. Time	Run Time	Mix	ppO2
45.0m	27.8	(30)	Air	0.21 - 1.16	45.0m	27.8	(30)	Air	0.21 - 1.16
24.0m	1.0	(33)	Air	1.16 - 0.71	24.0m	1.0	(33)	Air	1.16 - 0.71
21.0m	1.0	(34)	Air	0.71 - 0.65	21.0m	1.0	(34)	Air	0.71 - 0.65
18.0m	1.0	(35)	Air	0.65 - 0.59	18.0m	1.0	(35)	Air	0.65 - 0.59
15.0m	4.0	(39)	Air	0.59 - 0.52	15.0m	2.0	(37)	Nx60	0.59 - 1.50
12.0m	4.0	(43)	Air	0.52 - 0.46	12.0m	3.0	(40)	Nx60	1.50 - 1.32
9.0m	7.0	(50)	Air	0.46 - 0.40	9.0m	4.0	(44)	Nx60	1.32 - 1.14
6.0m	54.0	(104)	Air	0.40 - 0.34	6.0m	19.0	(63)	Nx60	1.14 - 0.96

OTU of this dive: 40
CNS total: 14.6%

OTU of this dive: 69
CNS total: 25.2%

4482.3 ltr Air ------> (6723.45 ltr for thirds)

2692.1 ltr Air ------> (4038.15 ltr for thirds)
756.8 ltr Nx60 ------> (1135.2 ltr for thirds)

Dive schedules demonstrate the advantage of accelerated decompression.

This significant reduction in decompression obligation and run time is definitely something worth reviewing.

All body tissues ongas during the descent and while diving at the maximum depth. As the diver ascends, depending on the compartment's degree of saturation, some tissues offgas and some others still ongas. Let's take a look at some numbers.

Depth		Run Time	Mix	Alveolar	Cpt ppN$_2$ (bar)		
(m)	(ft)	(min.)		ppN$_2$ (bar)	C1	C2	C7
45	150	30	Air	4.19	4.18	4.00	1.85
24	80	33	Air	2.59	3.55	3.62	1.92
21	70	34	Air	2.36	3.33	3.49	1.93
18	60	35	Air	2.13	3.19	3.37	1.93

Table 4.2: Partial pressures of nitrogen in selected compartments at different depths

The *alveolar* ppN$_2$ is calculated by multiplying the ambient pressure minus the water vapor pressure at the given depth by the fraction of nitrogen in the breathing mix. So ppN$_2$ at 45 meters (150 feet) = (1.01325 − 0.0493) * 0.79 * 5.5 = 4.19 bar approximately.

The numbers in bold indicate the highest ppN$_2$ at a given depth. We see clearly that as we start the ascent, the fastest compartment (C1, halftime = 5 minutes) was the most saturated compartment. After completing the 24-meter (80-foot) stop, the lead was transferred to the second fastest (C2, halftime = 8 minutes). On the other hand, the shaded numbers indicate that although the ambient pressure is reduced while ascending, the compartment is slow enough that its ppN$_2$ in still lower than the alveolar ppN$_2$. Until the 18-meter (60-foot) mark, the seventh-fastest compartment (C7, halftime = 54.3 minutes) was still ongassing.

The faster compartments obviously were more saturated than the slower ones. During the ascent, the ambient pressure was reduced, and the partial pressure of the inspired/alveolar nitrogen was accordingly reduced. These differences between the ppN$_2$ in the compartments and the ppN$_2$ in the alveolar gas are called the *pressure gradients* or the *inert gas gradients*. The term *diffusion gradients* describe the physical phenomenon more correctly, yet it's rarely used. In the faster compartments, these gradients are positive (compartment ppN$_2$ is bigger than alveolar ppN$_2$), so offgassing occurs. In some of the slower compartments the gradients are negative, so ongassing occurs, although the diver is heading toward shallower depths. As the various tissues ongas and offgas nitrogen at different rates while the diver ascends, the gradients keep changing. This means that for every segment of the ascent profile, there's a leading or controlling tissue compartment. This leading compartment has the highest nitrogen loading at this particular time segment of the ascent. Table 4.2 demonstrates that the 5-minute halftime compartment (C1) was controlling the ascent until the 24-meter (80-foot) mark, then the lead was transferred to the 8-minute one (C2).

In the air-only scenario — the 104-minute dive — reducing the ambient pressure (i.e., ascending) is the only way to create a gradient. However, we can't ascend too much because the drop in the ambient pressure influences bubble growth. In order to control bubble growth, every tissue tension must not exceed its M-value at the given depth.

In the air-EAN60 scenario, we breathe a lower ppN_2. This creates bigger gradients with no risk of bubble growth. The whole idea is that the M-values are linked to the tissue tensions and the ambient pressure, while the pressure gradients are linked to the tissue tensions and the partial pressure of the inspired/alveolar inert gas. By lowering the latter, we increase the pressure gradients without bearing the risk of exceeding the M-values.

For example, when breathing air at 15 meters (50 feet), the alveolar ppN_2 is approximately 1.9 bar. According to the numbers in Table 4.2, the seventh compartment would be almost at equilibrium, neither ongassing nor offgassing. If EAN60 is used, the alveolar ppN_2 would be approximately 0.96 bar; now the seventh compartment is offgassing.

The bigger gradients will also influence faster offgassing rates in the faster compartments. For example, the halftime of the fastest compartment (C1) is 5 minutes, which means that in case of offgassing, it will release half its tension in 5 minutes.

Now assume a 5-minute stop (not including the ascent time from the former stop depth) breathing air at 15 meters (50 feet). We know from Table 4.2 that the ppN_2 in this particular compartment at the end of the 18-meter (60-foot) stop was 3.07 bar. At the beginning of the 15-meter (50-foot) stop, the ppN_2 became 3.04 bar. The tension that this particular compartment can release in 5 minutes (its halftime) is 3.04 minus the alveolar ppN_2 (1.9 in case of air) divided by 2 = 0.57 bar.

For the very same 5-minute stop at 15 meters (50 feet) but breathing EAN60 instead of air, the tension that could be released is (3.04 – 0.96) / 2 = 1.04 bar.

In conclusion, it's evident that breathing lower inert gas content accelerates the decompression without bearing an additional risk of bubble growth. However, the elevated oxygen content requires strict tracking to mitigate the increasing risk of CNS oxygen toxicity. The current standard the majority of divers stick to is a maximum ppO_2 of 1.6 atm.[4]

Summary

In this chapter, we portrayed what happened when nitrox was introduced to the sports diving community in the early 1990s. We then explained what nitrox is and what its advantages are. Subsequently, planning dives using nitrox was illustrated, along with the accompanying threats. Finally, using nitrox for accelerating decompression and the significant reduction in decompression obligation were demonstrated.

CHAPTER 5
Mixed Gas

I n 1939 U.S. Navy divers working on salvaging the sunken submarine USS *Squalus* at a depth of 74 meters (243 feet) in the cold water of the North Atlantic reported dispersion of mind and loss of reasoning. Overseen by Albert Behnke (1903–1992), compressed air was replaced with a mixture of oxygen and helium called heliox, allowing the divers to avoid the cognitive impairment symptoms they experienced earlier. This confirmed Behnke's 1935 theory about the narcotic effect of elevated partial pressures of nitrogen.[1]

The USS Squalus *in dry dock after salvage (circa 1940)*
Photo courtesy of Wikipedia. Public Domain. "USS Squalus."

Which gas?

Divers using air or nitrox breathe a mixture of oxygen and nitrogen. Nitrogen has no particular benefit other than making up the gas mix to avoid oxygen toxicity. As depth increases, the drawbacks of breathing nitrogen become evident. Increased pressures of nitrogen lead to a state of euphoria and confusion commonly known as nitrogen narcosis. At elevated pressures, the dense nitrogen is not easily delivered through the regulator. This breathing resistance not only increases the onset and progression of narcosis but also creates turbulence in the airways of the body, which leads to reduced ventilation.

Now it's evident that if we want to go deeper, we have to reduce the amount of nitrogen in our breathing gas. We also have to reduce the amount of oxygen to avoid its poisonous effect at elevated pressures. To do that, we have to substitute some, or maybe all, of the nitrogen with a lighter, less narcotic gas. We have to substitute some of the oxygen as well, but we still need to leave a certain fraction capable of producing a ppO_2 of at least 0.16 bar at the minimum depth of usage, otherwise body metabolism will not be maintained properly, and hypoxia will occur.

A number of inert gas substitutes have been tested. An inert gas is defined as a gas that does not undergo chemical reactions under a set of given conditions. We know this set of conditions is fulfilled at sea level but not at elevated pressures, otherwise we would have not experienced the nitrogen narcosis problem at the first place. All mammals, including man, show disturbances at the level of the nervous system when exposed to increased breathing-mix pressures. These disturbances differ according to which gas is used. One of the most notable explanations is the Meyer-Overton correlation between lipid solubility and anesthetic potency. This correlation relates the narcotic effect of an inert gas to its solubility in the lipid (fatty tissues). In 1979 Mount suggested the narcotic effect to be determined by multiplying the solubility by the partition coefficient[2] (now known as partition constant, partition ratio, or distribution ratio).

Tests for a more weakly reacting gas than nitrogen at increased pressures have included argon, helium, hydrogen, and neon. Argon is denser than nitrogen, so it can't be used sufficiently through a regulator easily at depth. It is also more than twice as soluble in lipid as nitrogen, so it's more narcotic. However, some divers use a combination of argon and oxygen (frequently called argox) to accelerate decompression at shallow stops. Breathing argox starting at 15 meters (50 feet) up to 9 meters (30 feet) is believed to reduce the inert gas uptake, because argon is denser than the other inert gases, so it will diffuse in the tissues more slowly. Argon elimination from the body is supposed to be very slow as well, but for these short exposures the tissues would absorb very little argon. There's no empirical data to prove or deny these claims though.

Neon is much less soluble in lipid than nitrogen, so it is less narcotic, yet it's prohibitively expensive. On the contrary, hydrogen is inexpensive. It is less soluble in lipid, so it is less narcotic than nitrogen. One of its biggest advantages is that it's the lightest element known to man, so it is the best in terms of breathing ease at depth. Its lightness also makes it the "fastest" gas in terms of saturation and desaturation. The fundamental drawback associated with using hydrogen is that it becomes explosive and flammable when mixed with oxygen percentages over 4%. Out of these choices, helium is the least soluble in lipid, so it is the least narcotic. It is also sufficiently lighter than nitrogen.

On ascent, the ambient pressure is reduced, and offgassing starts to occur. The reduced inspired/alveolar partial pressure of the inert gas means that this gas diffuses back from the tissue to the blood. In the unfortunate situation of rapid ascent, bubbles may come out of solution. As the size of bubbles formed depends on the amount of dissolved gas, the higher the gas *solubility in blood,* the bigger the bubbles it promotes (higher solubility = more gas dissolved = bigger bubbles). This makes helium preferable to other inert gases. All in all, helium is considered the best all-around gas for deep diving. The high price of neon disqualifies it from practical consideration, thus making hydrogen the second-best gas for deep diving.

Gas	Molecular Weight***	Solubility in Lipid*	Solubility in Blood*	RNP**	Cost
Helium (He)	4.00	0.015	0.0087	4.26	Medium
Neon (Ne)	20.18	0.019	0.0093	3.58	High
Hydrogen (H₂)	2.02	0.036	0.0149	1.83	Low
Nitrogen (N₂)	28.01	0.067	0.0122	1.00	Low
Argon (Ar)	39.95	0.14	0.0260	0.43	Medium

*Solubility coefficients are quoted in *atm⁻⁽¹⁾*
**Relative narcotic potency; lower values for the more narcotic gases
***Molecular weights are quoted in *atomic mass units*

Table 5.1: The physical properties of various inert gases

Helium's thermal conductivity is considerably higher than that of nitrogen. It's a cold gas to breathe, which indicates that heat loss would occur more rapidly while breathing helium-based mixtures. The speed of sound in helium is almost three times the speed of sound in air. When helium is inhaled, there is a corresponding increase in the resonant frequencies of the vocal tract. That's why the voice of a diver breathing helium-based mixtures temporarily sounds high-pitched.

High-pressure nervous syndrome (HPNS)
On the basis of lipid solubility only, the narcotic effect of helium would manifest itself at around a 400-meter (1,300-foot) depth. However, in 1973 it was observed that pressure can reverse the effect of anesthetics and inert gases.[3] This pressure-reversal effect counteracts the weak solubility-in-fats potency, and some symptoms start to exhibit themselves as shallow as a 100-meter (330-foot) depth. These symptoms are coined in one term: high-pressure nervous syndrome (HPNS). As opposed to the euphoria and confusion of nitrogen narcosis, HPNS causes tremors, dysmetria, and muscle spasms. That's why breathing heliox is not preferred because the ppHe should not exceed a certain threshold. As a proposed solution to this problem, we add nitrogen to the breathing mixture. The helium-nitrogen-oxygen mix is called trimix. But adding some nitrogen is not enough. Blending the right mix for the dive according to the diver's physical capability is mandatory because HPNS incidents do happen.

On April 6, 1994, cave-diving pioneers Jim Bowden and Sheck Exley were supposed to dive to 305 meters (1,000 feet) at El Zacatón sinkhole in Mexico. Exley used trimix 6 (also known as heliair 6) as his bottom mix, Bowden used trimix 6.4. A term originally coined by Exley, heliair is helium topped off with air, so the ratio of oxygen to nitrogen is always 21:79. Exley's mix was 6% O_2, 22.6% N_2 and 71.4% He, or simply trimix (6, 71.4) as we would call it now. Bowden's mix was composed of 6.4% O_2, 24% N_2 and 69.6% He.

Bowden dived to 282 meters (925 feet), setting the then-new world record. Exley failed to return from his dive. Three days later his body was found with his computer reading 276 meters (906 feet). Gilliam, Exley's close friend, offers his opinion: *"The best educated guess would point to an HPNS incident. Exley had experienced a bad one in Africa that resulted in uncontrollable muscular spasms and multiple vision. (Exley pursued the exploration of a huge underwater cave at Bushmansgat, South Africa, diving to 263 meters (863 feet) in the fresh water of this system. During this dive, Sheck experienced visual, somatic and neurological symptoms of HPNS. The symptoms resolved during his ascent to his first deco stop at 120 meters (400 feet), and there were no persistent effects.) This may have manifested again with more violent tremors that could have triggered an oxygen convulsion or simply made it impossible to negotiate gas switches as necessary. His death will remain a mystery and a tragic loss to the cave community."*[4]

Using trimix

To mitigate the risk of getting incapacitated by narcosis, we can add helium to our breathing mix. But how much is appropriate?

To plan nitrox dives using air tables, we calculate the equivalent air depth (EAD), which is the depth that subjects the diver to the same partial pressure of nitrogen if the breathing mix were air. For example, the EAD of EAN40 at 25 meters (83 feet) is 16.6 meters (55 feet).

But how narcotic is EAN40 at 25 meters (83 feet)? Can we say that its equivalent narcotic depth (END) is 16.6 meters (55 feet)?

We used to believe so, although there was some evidence that oxygen plays a part in the narcotic effects of a breathing mixture. In 1978 a study concluded that a rise in ppO_2 to 1.65 ATA caused a significant decrement of 10% in mental function.[5] NOAA's diving manual states that oxygen has some narcotic properties and thus recommends treating oxygen and nitrogen as equally narcotic.[6]

So assuming only nitrogen is narcotic, then the END of EAN40 at 25 meters (83 feet) is 16.6 meters (55 feet). Assuming oxygen and nitrogen are equally narcotic, the END of EAN40 at 25 meters (83 feet) is 25 meters (83 feet). Fortunately, although helium has a relative narcotic potency, we consider it as non-narcotic.

To calculate the END of any trimix mixture at a given depth, multiply the absolute depth (the given depth plus the imaginary depth the atmospheric pressure exerts) by the oxygen plus nitrogen fractions in the mix, then subtract the depth the atmospheric pressure exerts. The general formula is:

$$\text{END (m)} = (1 - \text{fHe}) * (\text{Depth in meters} + 10) - 10, \text{ or}$$

$$\text{END (ft)} = (1 - \text{fHe}) * (\text{Depth in feet} + 33) - 33$$

For a 90-meter (300-foot) dive using a gas mixture containing 50% helium, the END would be 40 meters (130 feet).

In conclusion, to combat narcosis, considerable amounts of helium will need to be used. We discussed earlier that breathing helium under pressure can result in HPNS hits. Hence there is an inevitable trade-off between these two factors in choosing the right mix, and knowledge about how helium will behave in our bodies under pressure is compulsory.

Graham's Law states that the volume of gas diffusing into a liquid is inversely proportional to the square root of the molecular weight of the gas.[7] By applying this to inert gases diffusing into body tissues, we can get their saturation and desaturation speeds relative to each other. When helium was compared to nitrogen, the result was as follows:

Speed (He) / Speed (N$_2$) = SQRT (28.01) / SQRT (4.00) = 2.65 approximately.

This means that helium is almost 2.65 times faster than nitrogen. This is the diffusivity ratio, which we use to obtain the helium compartment halftimes in relation to those of nitrogen.

Let's take the most popular 5-minute compartment (halftime = 5 minutes) as an example. These 5 minutes are the time it takes the compartment to get half-filled (saturated) or half-emptied (desaturated), given that the gas in action is nitrogen. For the very same compartment, the helium halftime is 5 / 2.65 = 1.89 minutes approximately.

In conclusion, helium uptake and elimination can be tracked using the same method employed for nitrogen by dividing the set of halftimes by 2.65. The helium compartment will behave the same as the nitrogen compartment with an equivalent halftime. For example, the halftime of Bühlmann's compartment number 16 in his ZH-L16A algorithm is 635 minutes for nitrogen. For helium, the halftime of the very same compartment is 635 / 2.65 = 239.6 minutes approximately. This means that compartment number 16 in the helium set will behave more or less like compartment number 12 in the nitrogen set, as it has a very close halftime (239 minutes). Similarly, helium compartment numbers 14 and 15 behave in a virtually similar manner to nitrogen compartment numbers 10 and 11 respectively, as their halftimes are almost the same. Table 5.2, which is excerpted from Bühlmann's ZH-L16A algorithm, demonstrates that nitrogen calculations can be easily ported over to helium by utilizing the 2.65 scaling factor.[8]

Cpt #	N$_2$ Ht	a (N$_2$)	b (N$_2$)	He Ht	a (He)	b (He)
10	146	0.3798	0.9222	55.1	0.5256	0.8703
11	187	0.3497	0.9319	70.6	0.4840	0.8860
12	239	0.3223	0.9403	90.2	0.4460	0.8997
13	305	0.2971	0.9477	115.1	0.4112	0.9118
14	390	0.2737	0.9544	147.2	0.3788	0.9226
15	498	0.2523	0.9602	187.9	0.3492	0.9321
16	635	0.2327	0.9653	239.6	0.3220	0.9404

Table 5.2: An excerpt from Bühlmann's ZH-L16A algorithm with coefficients (a) and (b)

One of the key concepts Bühlmann demonstrated is that the tolerated partial pressures of two different gases in the same compartment will vary according to their solubility coefficients in the *transport medium* that delivered those gases to that compartment. The transport medium in our case is the blood plasma. The solubility in blood of nitrogen and helium are 0.0122 and 0.0087 atm^{-1} respectively, which means that the ratio between them is 0.0122 / 0.0087 = 1.4 approximately. Now let's divide the (a) coefficients of each compartment. For example, the last compartment (a) coefficients of helium and nitrogen are 0.3220 and 0.2327 respectively, which means that the ratio between them is 0.3220 / 0.2327= 1.38 approximately. In other words, lower helium solubility in the blood means higher tolerated partial pressure in the compartments. The (b) coefficient is the reciprocal of the slope of the M-value line (1 divided by the value of the slope), so varying solubility is irrelevant to it.

Intermediate critical tensions (M-values) can be calculated based on the ratio of nitrogen and helium in the array of tissue compartments. For every tissue compartment, coefficient (a) of the inert gas mix (N_2 + He) and coefficient (b) of the same mix are computed in accordance to the partial pressures of nitrogen (ppN_2) and helium (ppHe) as follows:

$$a\ (N_2 + He) = (a\ (N_2)\ ^*\ ppN_2 + a\ (He)\ ^*\ ppHe)\ /\ (ppN_2 + ppHe)$$

$$b\ (N_2 + He) = (b\ (N_2)\ ^*\ ppN_2 + b\ (He)\ ^*\ ppHe)\ /\ (ppN_2 + ppHe)$$

Now that the body tissues ongas helium considerably faster than nitrogen, it's axiomatic to expect that at the end of the bottom time helium loadings will be higher than nitrogen loadings. The exception would probably be the very fast tissue compartments. They will reach the full saturation level so the tissue loadings for both nitrogen and helium would be equal.

Actually this is not the exact case. Depending on the depth and the bottom time, the nitrogen loading of the slow compartments would be higher than the helium's. The reason is that before the dive starts (let's assume it's the first dive in several days to eliminate the effect of residual inert gas), all the tissues are supposedly at equilibrium with the alveolar inert gas at the surface (given that it's not a dive at altitude where the diver's body did not have sufficient time to acclimatize). More about that in Chapter 8. Since the percentage of helium in the air we breathe at the surface is negligible (practically zero), the tissues will be free of helium before the dive starts. This is not the case for nitrogen, as the atmospheric air includes a considerable amount of that gas.

Now let's compare the fastest and the slowest tissue compartment loadings before the descent and at the end of the bottom time. We'll also check on a couple of tissue compartments in between. Assuming a bottom mix with equal nitrogen and helium contents and with sufficient oxygen content to support life from the beginning of the dive, trimix (20, 40) would be perfect. This mix contains 20% oxygen, 40% nitrogen and 40% helium. Let's assume a dive to 60 meters (200 feet) for 45 minutes.

The ppN_2 before the dive is calculated by multiplying the atmospheric pressure minus the water vapor pressure by the fraction of nitrogen in the atmospheric air. So ppN_2 (at sea level before diving) = (1.01325 − 0.0493) * 0.79 = 0.76 bar approximately.

Cpt #	N₂ Ht	ppN₂ (bar)		He Ht	ppHe (bar)	
		Surface	Leaving bottom		Surface	Leaving bottom
1	5	0.76	2.70	1.88	0.0	2.70
5	27	0.76	2.09	10.21	0.0	2.62
12	239	0.76	0.99	90.34	0.0	0.79
16	635	0.76	0.85	240.03	0.0	0.33

Table 5.3: Some tissue loadings before the descent and at the end of the bottom time

Now we have the complete picture. The tissues were preloaded with 0.76 bar of nitrogen even before the descent starts. At the end of the bottom time, the helium loading managed to overtake the nitrogen's only in the faster compartments. Table 5.3 demonstrates this (the darker shade shows higher compartment loading).

The typical practice here is to use nitrox to accelerate the decompression. By switching to EAN40 at 30 meters (100 feet), the inspired/alveolar helium content becomes zero, which maximizes the helium gradient and thus accelerates helium offgassing to the max. On the other hand, the inspired nitrogen content is increased from 40% to 60%. This means that upon switching gases at 30 meters (100 feet), the inspired ppN_2 increased from 1.6 bar to 2.4 bar, thus the rate of nitrogen offgassing is expected to slow down. Actually nitrogen ongassing in some compartments would occur.

So the very same tissue could engage in more than one inert gas exchange, with inert gases going in and out at the same time. For example, it could offgas helium and ongas nitrogen. We can add some helium to the deco gas to decrease the nitrogen content — for example use trimix (40, 25) (that's 40% O₂ and 25% He) instead of EAN40. Will this decrease the total decompression obligation? It depends, mainly on the bottom time. In our example dive, using trimix (40, 25) instead of EAN40 would actually increase the total decompression obligation.

Helium is a fast gas, meaning that the fast compartments will be quickly saturated with it, sooner than on air dives anyway. Would that require deeper stops for trimix dives? Not necessarily. Helium ongases fast, yet offgases fast as well. So during the initial ascent phase, more helium will be eliminated than nitrogen. This fast offgassing during the initial ascent phase suppresses the need for deeper stops. In our example dive, using air as a bottom mix instead of trimix (20, 40) won't result in the stops starting any shallower.

By the end of the dive, the fastest compartments will be totally clear of helium. Upon surfacing, the nitrogen loading will be more than the helium loading for all the compartments.

In the sample dive schedule with inert gas loadings on the next page, you will notice that there's a "total loading" column. How can we add the nitrogen loading to the helium loading despite their different halftimes? Schreiner established an important concept: The total inert gas pressure (total load) in a given compartment is the sum of the partial pressures of all inert gases in that compartment, even if they have different halftimes. As we've seen, the halftimes of the helium were mathematically derived from those of the nitrogen according to the diffusivity ratio. So although the halftimes are different, each pair of halftimes (nitrogen and helium) constitutes a single hypothetical compartment.

Ultimate Planner 1.5 by Asser Salama.

Warning: This software is intended for demonstration purposes only. The author accepts absolutely no responsibility for the schedules generated by this software. Use it at your own risk.

Buhlmann-GF/U: OFF
Model: ZH-L16B
GF Low: 30% - GF High: 85%
Altitude: 0.0m
Leading compartment enters the decompression zone at 45.1m
Run time includes the ascent time required to reach the stop depth

Depth	Seg. Time	Run Time	Mix	ppO2
60.0m	42.0	(45)	Tx20/40	0.20 - 1.40
33.0m	1.0	(49)	Tx20/40	0.92 - 0.86
30.0m	2.0	(51)	Nx40	0.86 - 1.60
27.0m	1.0	(52)	Nx40	1.60 - 1.48
24.0m	2.0	(54)	Nx40	1.48 - 1.36
21.0m	3.0	(57)	Nx40	1.36 - 1.24
18.0m	4.0	(61)	Nx40	1.24 - 1.12
15.0m	6.0	(67)	Nx40	1.12 - 1.00
12.0m	8.0	(75)	Nx40	1.00 - 0.88
9.0m	10.0	(85)	Nx80	0.88 - 1.52
6.0m	47.0	(132)	Nx80	1.52 - 1.28

OTU of this dive: 190
CNS total: 78.9%

4931.0 ltr Tx20/40 ------> (7396.5 ltr for thirds)
1068.8 ltr Nx40 ------> (1603.2 ltr for thirds)
1427.3 ltr Nx80 ------> (2140.95 ltr for thirds)

Cpt. #	N2 Loading	He Loading	Total Loading
1	3.052m	0.0m	3.052m
2	3.16m	0.0m	3.16m
3	3.688m	0.0m	3.688m
4	4.896m	0.006m	4.902m
5	6.639m	0.087m	6.726m
6	8.265m	0.438m	8.703m
7	9.445m	1.28m	10.725m
8	10.022m	2.508m	12.531m
9	10.091m	3.67m	13.761m
10	9.899m	4.291m	14.189m
11	9.644m	4.483m	14.127m
12	9.36m	4.4m	13.76m
13	9.078m	4.116m	13.194m
14	8.815m	3.703m	12.517m
15	8.582m	3.235m	11.816m
16	8.382m	2.762m	11.144m

Complete dive schedule with inert gas loadings at the end of the dive

Isobaric counterdiffusion (ICD)

Like the other ingredients of the decompression theory, ICD (also abbreviated IBCD) is still not fully understood. Our current understanding is that it would occur at gas switches, when the diver swaps over from a mix containing a high percent of one inert gas to another mix containing a high percent of another inert gas. What happens in this situation is that one inert gas enters a particular tissue faster than the other leaves. Remember the very same tissue could engage in more than one inert gas exchange, with inert gases going in and out at the same time.

What we believe takes place is that nonmetabolic gases diffuse in opposite directions (in and out of the same tissue) according to their respective partial pressure gradients without being subjected to changes in the ambient pressure (hence the term *isobaric*, which means "same pressure"). As the tissue tension is the sum of all tensions exerted by the inert gases present in dissolved state,[9] the result is that the tissue tension will rise. If the tissue tension passes its critical supersaturation threshold, bubbles will form. This may occur most commonly in three situations.

The first situation that triggers an ICD hit is when the inert gas breathed by the diver is heavier than the inert gas surrounding the body. In this situation, transdermal diffusion takes place. This is often referred to as *superficial ICD,* and it could result in skin lesions similar to symptoms of skin bends. Powell explains: *"There is a phenomenon called 'isobaric counterdiffusion.' This occurs when a person [or animal] breathes nitrogen [a slowly diffusing gas] at pressure while surrounded by helium [a fast diffusing gas]. A change in ambient pressure is not required for copious venous bubbles to form. It is proof that in vivo micronuclei exist, and if exposed to supersaturation for an hour, visible gas bubbles form."* [10]

But how does that apply to us? We don't tend to immerse ourselves in helium. Well, don't use helium-based mix for drysuit inflation. Although it seems highly unlikely (helium-based mixes are more expensive than nitrox and are pretty bad as thermal insulators), I've seen divers do that, probably because of a more convenient hose routing (drysuit's inflator hose on back gas regulator).

The second situation is when the diver switches from a breathing mix containing a high content of a slow inert gas (nitrogen) to another mix containing a high content of a faster inert gas (helium). In this instance the helium enters the tissues faster than the slower nitrogen can leave. The fast influx of helium would cause higher inert gas tension than the tissues could tolerate. This is often referred to as *deep tissue ICD*. So switching back from a nitrox deco mix to a helium-rich back gas — probably as a maneuver to deal with a lost deco gas situation or as an air break to manage the O_2 clock (CNS status) and avoid oxygen poisoning (if you believe in that) — might lead to an ICD hit. Practically speaking, deep tissue ICD is not an issue for short dives. For longer dives, overcoming this situation is as simple as switching back to a leaner nitrox mix or to a travel gas in which the helium content is not as high as that in the back gas.

The third situation is the most controversial and the most dangerous. It involves the diver switching from a breathing mix containing a high content of a low-solubility inert gas (helium) to another mix containing a high content of a higher-solubility inert gas (nitrogen). It sounds a bit illogical that switching from high-nitrogen to high-helium mixes (the second situation) would cause ICD, yet switching from high-helium to high-nitrogen mixes (the opposite) would also cause ICD.

What we believe happens in the third example is that upon a switch from a high-helium to a high-nitrogen mix, the newly introduced nitrogen percent with its high solubility, will dissolve in the tissues fast enough that it would raise their tensions significantly. This sudden increase would result in the tissue tensions exceeding their M-values. Symptoms associated with this type of ICD are vomiting and extreme vertigo. These symptoms are almost identical to those associated with inner-ear infections, so this led us to assume that the place where this type of ICD takes place is the inner ear, particularly the vestibular apparatus. That's why the condition is often referred to as either *inner-ear decompression sickness* (IEDCS) or *vestibular bends*.

Our current understanding is that ICD would rarely manifest itself for dives shallower than the 80- to 90-meter (265- to 300 -foot) range. ICD hits at shallower dives are usually attributed to large patent foramen ovale (PFO) to be discussed in Chapter 8.[11] Although we still don't have the full concept of the ICD phenomenon (probably a series of related phenomena happening in a particular sequence), some computer decompression-planning tools warn us of the possibility of getting an ICD hit, particularly of the third, most controversial type. So what model do they use?

There is more than one approach, but the most commonly used is Stephen Burton's model, which aims to *not* increase the quantity of dissolved gas in the body tissues.[12]

Given the quantity of a particular dissolved gas in a saturated medium is equal to the solubility coefficient of this gas in that medium multiplied by the current saturation pressure, and the total quantity of dissolved gas in that medium is the sum of the quantity of each gas present, it's pretty easy to calculate the total quantity of dissolved gas at gas-switch stops and subsequently calculate the tissue tensions before and after the switch.

So at each gas-switch stop, the following condition is to be verified, otherwise an ICD warning message is displayed:

(Fraction of helium in next mix * Helium solubility + Fraction of nitrogen in next mix * Nitrogen solubility) <= (Fraction of helium in current mix * Helium solubility + Fraction of nitrogen in current mix * Nitrogen solubility)

The common practice, and what Burton endorses, is to use the solubility coefficients in lipid (fat), which are 0.015 and 0.067 atm$^{-(1)}$ for helium and nitrogen respectively. As you can see, nitrogen is almost 4.5 (0.067 / 0.015) times more soluble than helium in lipid. This means that by increasing the nitrogen percentage no more than ⅕ (a little bit more conservative than ¼.₅) of the reduction in the helium percentage, ICD hits could be avoided. This is what we know as the *rule of fifth*.

Now consider switching from one of my favorite gases, trimix (20, 25), to another gas I frequently use, EAN40, at 30 meters (100 feet). The *total amount of dissolved gas* before the switch was:

$$(0.25 * 4 * 0.015) + (0.55 * 4 * 0.067) = 0.015 + 0.1474 = 1.624 \text{ bar}$$

After the switch it became:

$$(0 * 4 * 0.015) + (0.6 * 4 * 0.067) = 0 + 0.1608 = 1.608 \text{ bar}$$

Since the total pressure exerted by the dissolved gases (total tissue tension) did not increase, this gas switch should be safe.

A simpler way to look at it is to employ the rule of fifth, which indicates the percentage of nitrogen is not to be increased by more than ⅕ the percentage of reduction in helium. When switching from trimix (20, 25) to EAN40, the increase in nitrogen is 5%, whereas the reduction in helium is 25%. This ratio (5 / 25) did not exceed ⅕, so this gas switch should be safe.

This common practice is at odds with the assumption that ICD occurs in the inner ear because the scientific community tends to model the inner ear as an aqueous tissue (a biological fluid close to water), not lipid.[13] The solubility coefficients of helium and nitrogen would accordingly be 0.01 and 0.015 atm[-1] respectively, which will obviously result in less-conservative ICD warnings if Burton's model is employed. On the other hand, both the diving and the scientific communities seem to agree on neglecting the diffusion effect. So far the models that aim to prevent ICD are solubility-only models. Some planning tools, such as Ultimate Planner, offer both options (aqueous versus lipid tissue), along with the capability of specifying at which depth you want the program to start simulating ICD.

Summary
In this chapter, we illustrated the inadequacy of using oxygen–nitrogen mixes for deep diving. The physical properties of various "inert" gases were compared, showing why helium is considered the best all-around gas for deep diving. Subsequently, we demonstrated why it's not appropriate to just replace all the nitrogen with helium and why using oxygen–nitrogen–helium mixes would be more favorable. With this added complexity, we portrayed how the tissues would respond to the simultaneous exposure of more than one nonmetabolic gas, what the involved risks are, and how to mitigate them.

CHAPTER 6
Dual-Phase (Bubble) Models

According to Dr. Michael Powell, the main hypothesis on which dissolved-gas models are based — that dissolved inert gases in the body tissues and in the blood will not form bubbles unless their respective M-values are exceeded — is flawed:

"Scientists have known for about two hundred years that simple supersaturation will not produce gas bubbles. This whole process is referred to as a phase change and covers both bubble formation and crystal growth, both from supersaturated conditions. The supersaturation limits for the production of a decompression gas phase in vitro in quiescent water is about 136 atm (2,000 psi). This exceeds by two orders of magnitude that supersaturation for in vivo systems. Traditionally, a change of 1 ATA (10 meters / 33 feet to the surface) is thought to be the decompression limit (for very long dives, anyway)."[1]

Asymptomatic (silent) bubbles
Powell continues:

"Within the last twenty years, the easy formation of decompression gas bubbles has been attributed to the presence of preformed tissue microbubbles capable of serving as 'seeds' upon which the decompression gas phase will grow during ascent. The postulation of the existence of these quasi-stable, preformed microbubbles in living tissue forms the basis for the decompression systems known as two phase.

"One origin of the microbubbles in physical models is physical stress, as when surfaces with water in between are separated (processes called by names such as tribo-nucleation, viscous adhesion or tacky adhesion). In living tissue, movement of muscles affords opportunities for surfaces to separate. This is especially true with the walls of capillaries. The physicist Edmond Newton Harvey (1887-1959) proposed in the early 1940s that these microbubbles

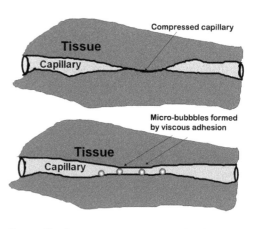

could reside between the walls of capillary cells; these were sometimes called 'Harvey pores.' This whole process is referred to as 'stress-assisted nucleation.' These processes allow bubble formation to arise since work is added to the system and bubbles are not formed solely by gas supersaturation.

"The origin of these tissue gas microbubbles have been attributed to a gas phase in either hydrophobic (water-resisting) cavities, surfactant-stabilized microbubbles, or,

as discussed here, arising from musculoskeletal activity. Experiments with rats, frogs and crabs showed that physical activity increased decompression bubble formation. These studies were also performed with men as test subjects. It is this last category that I discussed under stress-assisted nucleation. The lifetimes of these micronuclei have been variously estimated from as short as a few days to several weeks.

"It was my idea that bubbles of a radius that could play a role in DCS lived but a few hours. A test was made at NASA to estimate this lifetime. After a prescribed regimen of exercise and rest, a depressurization was made and gas bubbles were detected by precordial Doppler monitoring. In a crossover design, twenty individuals (15 men and 5 women) at sea level exercised by performing squats (150 knee flexes over ten minutes) either at [a] the beginning, [b] middle, or [c] the end of a two-hour chair rest period. There was no oxygen prebreathe. Seated subjects were then depressurized to 6.2 psi (6,700 meters/ 22,000 feet) for 120 minutes with no exercise performed at altitude. Ten of twenty subjects with Doppler-detectable bubbles in the pulmonary artery demonstrated greater bubble incidence with exercise performed just prior to depressurization, with decreasing Spencer bubble grades and incidence as the duration of rest increased prior to depressurization. The other 10 subjects never produced any detectable bubbles. They were resistant even with exercise. An analysis of the Doppler bubbles by a summation technique (Doppler gas volume) yielded the average for the subjects and definitely showed more bubbles with a short resting duration.

*"Analysis indicated that the micronuclei producing the Doppler bubbles had an average half-life of approximately 60 minutes under these conditions. Some subjects had longer halftimes and some shorter. **For recreational divers, this would indicate that nuclei do not persist for long times, but it must be remembered that activities such as walking constantly produce new ones. The worst offenders would be, e.g., heavy lifting, climbing ladders and surface swims.***

"An interesting small study was begun while I was at NASA, and I understand that interest remains even years after I have left. This is the question of sizing the tissue micronuclei. The basic device was developed for NASA by Creare Inc. when I was the contract monitor. They developed a bubble-detection-and-sizing instrument using a dual-frequency ultrasound device that emitted 'pump' and 'detection' ultrasound signals at two frequencies. The low-frequency pump signal caused a bubble of a certain radius to resonate. When the first frequency hit a resonating bubble, mixing signals are returned at the detector as the sum and difference of the two frequencies. Another transducer detected these. A graduate student used this as a part of his thesis research and was able to show, to a limited extent, that bubbles in gelatin and small animals could be sized, and these corresponded to visual images in a microscope. Further funding limited these studies. This work could have contributed to our understanding of tissue nuclei, but, alas, money is often an issue."

In the 1970s, Doppler technology was incorporated into decompression research. This then-new technology allowed researchers to measure the existence of bubbles in the diver's body. The results clearly demonstrated that bubbles form after almost every dive, even if the diver is not experiencing any problems. But are these silent bubbles bad? Do they have long-term effects? In 1997 a study suggested that they could produce subclinical brain lesions in divers with no history of DCS.[2] In 2009 a study evaluated 113 asymptomatic male military divers and 65 nondiving males in good health for cerebral white-matter lesions using magnetic resonance imaging (MRI). MRI revealed brain lesions in 26 of the 113 divers (23%) and in seven of the 65 nondivers (11%). The difference was statistically significant to confirm the assumption that cumulative, subclinical injury to the neurological system may affect the long-term health of military and recreational divers.[3]

The evolution of dual-phase (bubble) models

The existence of asymptomatic bubbles influenced decompression researchers to develop explicit physical models for bubble nucleation in supersaturated fluids. The development of these bubble models would permit the computation of decompression schedules based entirely on established physical principles. These dual-phase models will not only account for micronuclei existence but will also not discard the essential supersaturation part. As Powell said:

"Make no mistake — supersaturation is essential. You will form many micronuclei in the gym while lifting weights, but you will never get decompression sickness.

"For scuba divers, nucleation refers to a step in the formation of the gas phase in the body that eventually can result in decompression sickness. Divers have been taught that decompression bubbles form with ascent and the attendant decrease in ambient pressure. This is not actually correct — tiny microbubbles are there prior to depressurization. While decompression sickness cannot result without supersaturation, it also cannot occur in the absence of micronuclei."[4]

In the 1960s Hugh LeMessurier (1912-1976) and Brian Hills (1934-2006) took instrument recordings of the decompression schedules routinely followed by pearl divers, particularly the Okinawans who were operating in the Torres Strait and elsewhere along the northern coast of Australia. The remarkable practice of those divers was derived over a century (between 1850 and 1950), purely by trial and error at the expense of about 4,000 lives and many more serious injuries. Hills estimated this vast distillate of human experience to be more than 100 million air dives, ranging from 30 meters (100 feet) to 90 meters (300 feet) and up to 2 hours bottom time. They even included some repetitive dives. The findings were published in 1965. At that time, it was the researchers' conclusion that the data they collected could only be explained on a dual-phase model they called *thermodynamic approach*.

In 1966 Hills described this work in greater detail as a part of his seminal doctoral thesis submitted to the Faculty of Engineering, University of Adelaide, Australia. Hills illustrated that those divers were using a very efficient decompression practice. Based on the data in hand, he further developed the model and called it *thermodynamic and kinetic approach*. Although their methods were developed empirically, the number of successful dives (without DCS symptoms manifesting themselves) carried out was fair enough to illustrate that the results of employing these methods are statistically significant, thus should not be dismissed. It was astonishing that their practice was not only successful but also more economical on time than the Navy's.[5]

At that time, the neo-Haldanean calculation methods were at their peak. In comparison to them, the pearl divers spent much more time deeper at the start of decompression.

This discovery influenced much scientific work, leading to suggesting that an equilibrium state rather than a supersaturated state would be more relevant in determining the onset boundaries of DCS.

Hills built his model on only one tissue type, claiming there was only one tissue type where the onset of DCS took place (typically the limbs). So if we could prevent the onset of DCS in the limbs, all types of DCS would be subsequently controlled. Hills was not the first to regard the body as a single unit. In 1937 Behnke mentioned that, with the exception of organ tissues of high fat content (bone marrow, spinal cord and myelin sheaths), the division of the body into tissues that saturate and desaturate at different rates was largely arbitrary.[6]

Hills also assumed this sole, critical tissue consisted of millions of "microregions" and there was a reservoir of nuclei in it. Although most areas of this tissue would retain their supersaturation, it took only one microregion to dump its gas for DCS to occur. He called this scenario the *worst possible case* and considered it the most relevant. Once this "gas dumping" started, the tissue's dissolved gas pressure could decrease and the pressure inside the receiving nucleus would increase, forming a stable gas bubble. When the two pressures became equal, phase equilibrium occurred. Now only *inherent unsatuation* (also known as *oxygen window*, to be discussed in Chapter 8) could eliminate this bubble, as it caused a partial pressure difference of inert gas between the inside and outside of the bubble. Other researchers called this approach *nil supersaturation* and *zero supersaturation*. Hills frequently referred to it as *preventive decompression*.

Unfortunately, Hills' model didn't enjoy much acceptance. While it did pioneer the concept of deep stops (which was already introduced — empirically — into some commercial diving schedules), Hills' thermodynamic theory was such a radical innovation that the traditional minds of that time failed to realize the shallow stops were less in time than those advised by neo-Haldanean models, which suggested these traditional dissolved-gas models followed a "bend then mend" approach.

One day Hills observed two bubbles in a living tendon, one shrinking and the other growing — simultaneously. Intermittent perfusion was discovered later, but that made mathematical modeling next to impossible. At the time (circa 1980), a dual-phase model involved mathematics that were too complex to program into personal computers.

Hills finally gave up after 20 years of advocating dual-phase models. Years later, another dual-phase model emerged.

Varying permeability model (VPM)

Chronic conditions such as dysbaric osteonecrosis (also known as aseptic bone necrosis, a destruction of bone tissues) were historically associated with bubble formation during diving. Studies in the 1970s concluded the risk of aseptic bone necrosis was 10 times greater in navy divers and 100 times greater in fishermen in comparison with the risk in the general population. These findings supported the existing belief that the neo-Haldanean models in use at the time follow a "bend then mend" approach. Several studies to investigate the bubble-formation and growth mechanisms took place, one of which was conducted at the University of Hawaii by David Yount (1935-2000) and other researchers. This group later became known as the "tiny bubble group."

Unlike its predecessor, VPM retains the established Haldanean practice of characterizing the body by a number of tissue time constants (tissue compartments). However, the conventional M-value calculation (pressure reduction limits) is replaced by a much more complex computational algorithm to trace the bubble nuclei through the pressure history. It also deals with the effects of metabolism by considering the pressure due to oxygen, carbon dioxide, and water vapor. Although modeled as a single constant, the pressure of these noninert gases can alter the pressure of the gases inside the bubble, which would lead to some effects such as *inherent unsatuation* and the possibility of *oxygen bends* (discussed in Chapter 8). Like Hill's thermodynamic model, VPM introduces much deeper stops than those advised by neo-Haldanean approaches to staging divers.

Just like any neo-Haldanean model, VPM uses a spectrum of hypothetical tissue compartment halftimes for each inert gas the diver uses. Most software developers incorporating VPM in their decompression programs use Bühlmann's ZH-L16 pair of halftime sets (one for nitrogen and the other for helium). These sets simulate the inert gas uptake and elimination throughout the dive. The Schreiner equation calculates the gas loadings during the descent and ascent segments, whereas the Haldane equation calculates the gas loadings during the constant depth segments.

To stage the diver during the ascent to the surface, neo-Haldanean models use linear pressure reduction ratios (typically known as M-values) the diver shouldn't exceed. VPM does not abide to that. Instead, it inserts a hypothetical bubble "seed" into each tissue compartment. Throughout the dive, the dissolved gas is tracked via the tissue compartments, and the free gas is tracked in and out of the compartment bubble seeds. The diver is then staged through chasing the pressure difference between the tissue compartment (tissue tension) and the pressure inside the bubble seed (bubble pressure).

To describe the bubble behavior, the tiny bubble group conducted several in vitro experiments. Although the body tissues are composed of aqueous and lipid substances, the experiments were conducted on gelatin rather than water or oil. The advantage of gelatin over water and oil is that it traps any bubbles that appear during decompression and doesn't allow them to ascend to the surface. This way the bubbles can be observed and counted. The tiny bubble group published scores of research papers on bubble formation, nucleation in supersaturated fluids, isobaric growth and skins of varying permeability. Then they extended their research and published another series of research papers on the evolution, generation, regeneration and determination of the radii of gas cavitation nuclei. Let's take a glance at what they deduced.

At equilibrium, the bubble seed, or simply bubble, neither grows nor shrinks. To achieve this state, the pressure inside the bubble must be equal to the pressure outside the bubble. Right? Well, not exactly.

There is another pressure we have to consider: the pressure due to surface tension, which is the pressure that causes the bubble to maintain its spherical shape. Powell explains: *"All bubbles have a constricting force known as surface tension. This is produced by the unbalanced forces on water molecules when they are not completely surrounded. If one side is against a gas (as with a bubble), the intermolecular forces are unbalanced. The water molecules tend to pull together, and the bubble is constricted. This inward pressure from surface tension is referred to as the Laplace pressure (named after the French physicist Pierre-Simon, marquis de Laplace, who first described it in the early 1800s). As bubbles become smaller, the Laplace pressure becomes hundreds of atmospheres, and bubble growth is impossible."* [7]

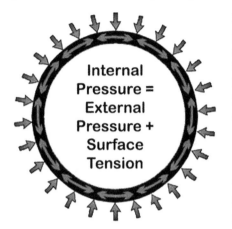

Now we have three pressures: the ambient pressure (outside the bubble), Laplace pressure or the surface tension (holding the bubble together), and the bubble pressure (inside the bubble). For the bubble to stay in equilibrium, the bubble pressure must be equal to the ambient pressure plus the surface tension.

P (Bubble) = P (Ambient) + P (Laplace)
(6.1)

The surface tension, or Laplace pressure, is *inversely* proportional to the radius (or size) of the bubble.

P (Laplace) = 2 * Gamma / R   (6.2)

Where gamma is the surface tension constant, and R is the radius of the bubble.

By solving these two equations together we conclude that:

R = 2 * Gamma / P (Bubble) – P (Ambient)   (6.3)

From the last equation we clearly see that the smaller the bubble size (radius), the bigger the pressure inside the bubble. Now let's revisit the last sentence of Powell's statement: *"As bubbles become smaller, the Laplace pressure becomes hundreds of atmospheres, and bubble growth is impossible."*

So far so good. The bubble tends to collapse, or at least not grow, thus should cause no problems. Unfortunately, the situation is much more complicated. We have inert gases, both in the bubbles and in solution (in the surrounding tissues, and remember that we have an array of tissues defined by a spectrum of halftimes). During the descent and the bottom-time phases of the dive, the inert gas in solution (tissue tension) increases with time (ongassing). Because we have varying uptake rates, the tissues will be loaded with various levels of dissolved gas. Now if the tissue tension is higher than the pressure inside the bubble, the inert gas will *diffuse* from solution into the bubble, causing it to grow. In layman's terms, **the bubble is not exactly a balloon.**

To prevent the bubble from growing, we need to make sure that the tissue tension does not exceed the pressure inside the bubble. So the *bubble no-growth equation* is defined as:

P (inert gas dissolved in solution) <= P (Bubble)   (6.4)

Or simply put:

Maximum allowed **pressure of dissolved inert gas = P (Bubble)   (6.5)**

The dissolved inert gas in solution is simply the partial pressure of inert gas in any given tissue compartment. By substituting in equation (6.3) for an inert gas (nitrogen for instance) we get:

$$R(N_2) = 2 * Gamma / ppN_2 - P(Ambient) \quad (6.6)$$

Where $R(N_2)$ is the radius of the nitrogen bubble, and ppN_2 is the maximum allowed partial pressure of nitrogen in solution.

From this we deduce that there is a *critical radius*. For bubbles with larger radii, the pressure inside the bubble will be less than the pressure of the inert gas in the surrounding tissue. So the gas will diffuse from solution into the bubble, causing it to grow. On the other hand, bubbles with smaller radii than the critical will never grow.

The other extremely important deduction is that now we have a formula to relate the *maximum allowed supersaturation*, which is [ppN_2 - P (Ambient)], to the critical radius $R(N_2)$.

Now that we know that the bubble is not exactly a balloon and that the gas will diffuse into the bigger bubbles, it's clear that the skin of the bubble is permeable. Experiments on gelatin exhibited that this permeability is only within a certain region. During the compression stage (descent), the bubble skin is permeable up to around an 82-meter (270-foot) depth. After that pressure, the bubble skin is impermeable and no diffusion takes place, hence the name *varying permeability*. Clearly, the calculation formulas for different regions are not the same.

One might think that there's no diffusion taking place and that the increase in bubble size is due to the pressure reduction on ascent. While Boyle's Law comes into play on ascent, the reduction in pressure is by no means sufficient to cause the increase in bubble volume witnessed during the experiments. According to Boyle's Law only with no diffusion taking place, it would require a container (test tube) 10 meters (33 feet) long to cause the bubble size to double.

During the descent, VPM computes the effective "crushing pressure" in each compartment as a result of compression. The crushing pressure is the difference in pressure between the ambient pressure and the pressure inside the bubble. This gradient acts to shrink the bubble smaller than its initial size at the surface. This phenomenon is of particular importance because the smaller the radius of the bubble, the greater the allowable supersaturation gradient upon ascent. Gas uptake during descent, especially in the fast compartments, will reduce the magnitude of the crushing pressure. The calculation of crushing pressures differs depending on whether the gradient is in the permeable region (gas can diffuse across the skin of the bubble) or not (molecules in the skin of the bubble are squeezed together so tightly that the gas can no longer diffuse in or out of the bubble). From this we conclude that, unlike neo-Haldanean models, VPM is more conservative when the descent rate is slower, because the effective crushing pressure is reduced.

Another extremely important note is that the crushing pressure in VPM is not cumulative over a multilevel descent. The crushing pressure is simply the maximum value obtained in any one discrete segment of the overall descent. Consequently, VPM must compute and store the maximum crushing pressure for each compartment that was obtained across all segments of the descent profile. In other words, a dive that stops initially at 60 meters (200 feet) then descends further to 100 meters (330 feet) results in less crushing pressures than a dive with an immediate descent to 100 meters (330 feet). That's why recompression (also known as sawtooth) profiles should be avoided or at least handled more conservatively if VPM is to be used for scheduling dive excursions. This is discussed in Chapter 8.

Now that we're done with the descent, the bottom time segment will affect neither the bubble size nor the crushing pressure. On the ascent, the bubble doesn't simply restore its original size. As we already know from Equation 6.4, unless the tissue tension exceeds the pressure inside the bubble during ascent, the bubble will not grow. VPM tracks the pressure inside the bubble during the ascent and stages the diver whenever a compartment's tissue tension is about to exceed the compartment's bubble pressure. In other words, **VPM dynamically generates its own "M-values," which we will call "VPM-values," throughout the ascent**. These VPM-values stage the diver by limiting the dissolved gas tension but in accordance to the compartment bubble pressure rather than by following some empirically derived values like ZH-L. However, as the main concept of staging divers by limiting the dissolved gas tension is still valid, switching to lower inert gas content (a richer deco mix) during the ascent would accelerate the decompression.

VPM used default initial radii values between 1 and 1.2 micron for nitrogen and 0.8 micron for helium. These values were set so that a *fully saturated* diver would have allowable supersaturation pressure that was consistent with the experimentally determined maximum allowable saturation depth for direct ascent to the surface.[8] Since the saturation depth of helium is bigger than that of nitrogen, the *initial critical radius* of helium is smaller than that of nitrogen. Separate crushing pressures are tracked for nitrogen and helium because they would have different critical radii.

The initial critical radii were derived experimentally for full saturation situations had an unacceptable outcome. Full saturation is when the diver stays at a particular depth long enough so that all the tissue compartments become fully saturated, a typical situation in the commercial diving world when underwater constructions or salvage operations that require extended work periods are considered. The expected yet unacceptable outcome is that the schedules generated by VPM for subsaturation (also known as bounce) dives, which is the type we sport divers are interested in, were very conservative.

Critical volume algorithm (CVA)
According to Doppler tests, asymptomatic bubbles exist in the diver's body after almost all dives, which clearly implies that the human body can tolerate *bubble volume* to a certain limit. Now VPM looks at incorporating multiple bubble behavior in the model. In other words, instead of controlling the size of the bubble only, let's also consider a relaxed *total volume of gas accumulated in the bubbles*.

In 1977 Tom Hennessy and Val Hempleman examined the feasibility of using the concept of critical released gas volume in DCS.[9] David Yount and Don Hoffman incorporated this concept and used a critical volume hypothesis, thereby assuming that signs or symptoms of DCS will appear whenever the total volume accumulated in the gas phase exceeds a designated *critical volume parameter*. This parameter was determined by making the total decompression times produced by VPM resemble those in the Tektite saturation dive and in the U.S. Navy (USN) and Royal Naval Physiological Laboratory (RNPL) manuals.[10]

VPM processes a complete deco schedule based on the *initial allowable supersaturation gradients*. CVA then relaxes these gradients and generates another schedule. The new schedule may cause the critical volume parameter to be exceeded. In that case CVA runs yet another profile. This back-and-forth search process continues until CVA hits its critical volume target, and that's why VPM is described as an iterative algorithm. The final schedule is now based on the *adjusted allowable supersaturation gradients*.

In 1990 Dr. Eric Maiken joined the VPM project. Throughout the 1990s he contributed several early versions of VPM desktop programs in BASIC, VBA and Mathematica. The latest version of his programs was nitrox, trimix, and multilevel capable. In 1998 Erik Baker, the developer of the gradient factors algorithm, joined the team. In 1999 Yount, Maiken and Baker collaborated to extend VPM to model repetitive dives. Although Yount died in April 2000, the core of the algorithm was finalized several months later. Baker's Fortran source code was released in the public domain and was dedicated in remembrance of Yount. This free availability of intellectual content encouraged software developers to incorporate the model in their desktop decompression-planning tools, which in turn increased the model's popularity. The initial feedback was fine, but as divers tended to go deeper, it became obvious that the model was flawed.

VPM with Boyle's Law compensation (VPM-B)
By investigating what's going on with the model beyond certain ranges, the project team discovered VPM does not account for bubble size changes during the ascent. In other words, the maximum allowable supersaturation gradients remained fixed throughout the deco schedule iterations and were only modified by CVA for subsequent deco schedule iterations. They did not change with the expected change in bubble size during the ascent, which made them invalid beyond the region they were calculated. In 2002 Baker fixed that and introduced a revised version. This revision incorporated Boyle's Law to compensate for bubble expansion during the ascent. As this added-on algorithm let the nuclei grow at each stop, it needed to reduce the initial critical radii and the critical volume parameter. The new default values were 0.55 micron for nitrogen (instead of 1-1.2) and 0.45 for helium (instead of 0.8). The critical volume parameter was dialed down from 2,286 meters/min (7,500 feet/min) to 1,981 meters/min (6,500 feet/min). The method of calculating the ascent ceilings and how they propagated through the ascent was also altered, as the increase in bubble size forced VPM to reduce the allowable gradients for the subsequent parts of the ascent profile, which in turn tended to increase the length of the shallower stops, especially when it came to deeper dives.

This revision was called VPM-B, and it became the standard version. The older VPM version (without Boyle's Law compensation) was frequently referred to as either VPM-0 or VPM-A. Implementing Boyle's Law compensation algorithm and the subsequent modifications were to fix a flaw in the model, not to add conservatism.

Now let's compare the output of VPM-B to that of ZH-L16B. Consider a 30-minute dive to 70 meters (230 feet) on trimix (17, 40), EAN40 and EAN80 with a descent rate of 20 meters/min (66 feet/min) and an ascent rate of 9 meters/min (30 feet/min). To match the total run time of the VPM-B generated schedule (96 minutes), we'll use gradient factors 50/95. The first required stop on VPM-B was at 42 meters (140 feet) versus 33 meters (110 feet) on ZH-L. Upon surfacing, the tissue loadings would be as follows:

Cpt #	VPM-B			ZH-L16B with GF 50/95		
	N₂ Load	He Load	Total	N₂ Load	He Load	Total
1	3.13m	0.0m	3.13m	3.089m	0.0m	3.089m
2	3.674m	0.0m	3.674m	3.474m	0.0m	3.474m
3	5.193m	0.005m	5.198m	4.732m	0.003m	4.736m
4	7.337m	0.081m	7.418m	6.676m	0.061m	6.737m
5	9.444m	0.489m	9.932m	8.696m	0.399m	9.095m
6	10.814m	1.477m	12.261m	10.084m	1.24m	11.324m
7	11.414m	2.872m	14.286m	10.766m	2.534m	13.3m
8	11.371m	4.255m	15.626m	10.833m	3.823m	14.656m
9	10.94m	5.049m	16.033m	10.513m	4.632m	15.145m
10	10.447m	5.26m	15.707m	10.105m	4.815m	14.92m
11	10.013m	5.069m	15.082m	9.734m	4.659m	14.393m
12	9.605m	4.671m	14.276m	9.379m	4.308m	13.687m
13	9.24m	4.159m	13.399m	9.058m	3.844m	12.902m
14	8.92m	3.598m	12.518m	8.774m	3.332m	12.107m
15	8.65m	3.048m	11.699m	8.534m	2.827m	11.361m
16	8.427m	2.542m	10.969m	8.335m	2.36m	10.695m

Table 6.1: Using Ultimate Planner's "Display tissue loadings upon surfacing" option to compare VPM-B to ZH-L16B

Upon surfacing, all the compartments of the schedule generated for VPM-B have higher loadings than those of ZH-L (higher values shaded). The only exceptions are the two fastest helium compartments. They are clear either way.

VPM-B conservatism

Now that we know that VPM-B assumes two initial critical radii, one for nitrogen and the other for helium, and only bubbles with bigger radii than these critical values are allowed to grow, VPM-B-based programs handle conservatism by assuming bigger initial critical radii values. For example, 5% conservatism is 0.55 * 1.05 = 0.5775 micron for nitrogen and 0.45 * 1.05 = 0.4725 micron for helium. Now VPM-B will need to control more bubble sizes to not allow them to grow, which in turn will result in more conservative schedules.

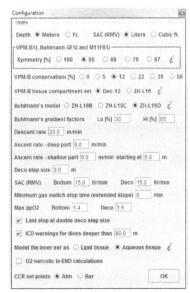

Ultimate Planner's configuration

For example, using VPM-B with 5% conservatism would result in a total run time of 98 minutes (instead of 96 minutes) for the same 30-minute, 70-meter (230-foot) dive. If using a conservatism level of 22%, the total run time would be 106 minutes.

VPM-B variations

In 2005 Ross Hemingway introduced a new model variation called VPM-B/E. In 2011 Shearwater introduced their VPM-B/GFS, and I introduced the VPM-B/U. /E stands for extreme, /GFS stands for gradient factor surfacing, and /U stands for ultimate. The three variations aim to generate more conservative schedules. I interviewed Hemingway to understand the logic behind each model.[11]

Ross, what can you tell us about your /E model variation?

"My method is proprietary. All I will say is that it looks at the internals of VPM and extends them in a manner when the dive's conditions within VPM become significant. The resulting extension time is in proportion to the underlying VPM-B model. The /E variation only starts to take effect when the decompression loading becomes large, usually affecting dives with 90-100 minutes or more total time.

"Interesting to note, most OC dives cannot experience any real change from VPM-B/E, because 100 minutes dive time is about the max for tank gas volume with reserves. Only at this point do the extended methods begin to outgrow the underlying VPM-B. For most divers, VPM-B and VPM-B/E are the same thing, simply because their dive isn't big enough to trigger the extra time from a /E plan.

"Also note that up to about 80 minutes, VPM-B is longer than ZH-L."

So why did you create it at the first place?

"I created VPM-B/E for Dave Shaw and his second 270-meter (886-foot) dive in the Boesmansgat cave. His first dive in that cave at 270 meters (886 feet) was with VPM-B and lots of padded extra time. For the second dive he wanted something in VPM-B, but longer, to plan the dive with. VPM-B/E was created, and the second dive was planned with VPM-B/E. But he never used it because he died on that next dive in January 2005 from a CO_2 hit while at 270 meters (886 feet)."

What about the /GFS? Is it all about comparing the VPM-B-generated profile to that of ZH-L and using the more conservative output?

"The new /GFS idea of Shearwater is a combined method. They look at two plans concurrently and then take the longer time frame from each. For the 100- to 200-minute dive time frame, it's about the same as the /E plan. But after 200 minutes or so, /GFS keeps growing and growing out of proportion and into silly numbers. This of course reflects the underlying problems of ZH-L. The failure of this two-model approach is that it only works by coincidence. It gives meaningful info in a certain region only and goes out of proportion beyond that."

What do you think of the /U?

"Your /U method theory — dissimilar gas rates — is already accounted for in the standard Haldane and Schreiner equations. So fiddling the offgas rates is a fudge — and a baseless one, too, with no calibrations to back it up. Furthermore, it interferes with the base calibrations for the model."

Ross, here's the Haldane equation:

Haldane Equation = Initial Gas Pressure + (Alveolar Gas Pressure – Initial Gas Pressure) * (1.0 – EXP (–Gas Time Constant * Interval Time))

Can you explain how it accounts for dissimilar gas rates?

"The equation has bearing of the gas direction (+/–). The result is relative to the inspired and existing pressures. That's all it needs. To try to force it to a bias based on direction is a fudge, a silly fudge."

I don't force it based on direction. As you said, that would be silly. And I think it would probably create more problems when the Schreiner equation is involved. However, the fact that it has a bearing and the result is relative to both the inspired and existing pressures do not mean, in my opinion, that it accounts for asymmetric gas kinetics.

"There is some testing in hot/cold changes in the dive. Getting cold slows down offgas rates — that's established. And the U.S. Navy Experimental Diving Unit (NEDU) uses this fact in its own dive stress testing by making the diver cold on the bottom to enhance the DCS rate.

"So I think your /U data is really just the hot/cold problem."

Do you want to add anything, Ross?

"I don't think any of them are necessary — 96% of all dives are less than 100 minutes, which means they were all diving base VPM-B plans regardless of the model choice they make. Only 0.5% of the dives are more than 200 minutes long, where real time differences can be seen in plan times. For these divers, how can you tell the difference between extra time added for safety and required extra deco time? You can't tell the difference. Adding extra time with a successful result does not imply it was needed time. And of course common sense says add more safety to big dives.

"The problem is that the frequency of 200-minute+ dives is pretty rare. Consider that if we all did 200 minutes every weekend, we would all get a better feel for the situation. I believe that divers would then trim off all the excess deco time that these extended methods present, and we would be back to the baseline times that regular VPM-B presents. But until then, extra safety is a good measure for success.

"So in conclusion, I don't think any of these add-on methods are really needed. All of them only take effect on really long dives. They are added mostly for extra safety reasons on big dives, where extra safety is the right thing to do."

As the /U simulates asymmetric gas kinetics, its use is not restricted to VPM-B. It can be added on to virtually any decompression algorithm for extra safety. One of the most common applications is use with the raw Bühlmann ZH-L16 model (without deep stops or gradient factors) for planning bailout for closed-circuit rebreathers (CCRs). This topic will be discussed in Chapter 8.

Reduced gradient bubble model (RGBM)

During the early and mid-1990s, Dr. Bruce Wienke extended the VPM to include repetitive and multiday diving in his proprietary RGBM, as this was not yet done by the VPM project team. Several versions of this model were released. Some dive computers and desktop planning tools incorporated a modified version, which is RGBM folded over a neo-Haldanean model. The idea is to modify the underlying dissolved-gas model by embedding bubble-reduction factors to account for reverse profiles, repetitive dives, and multiday diving.

Regarding the reverse profiles, RGBM assumes diving deeper than the previous dive stimulates previously "crushed" micronuclei into growth. RGBM considered the depth excess of the last dive compared with the one before it when calculating future decompression obligations. This assumption was the motive that made Yount and Wienke vote against the complete retraction of warnings against doing reverse profile dives in the 1999 conference.[12] They managed to get this relaxation conditioned by a maximum depth differential of 12 meters (40 feet). The final conclusion stated the workshop found no reason for the diving communities to prohibit reverse dive profiles for no-stop dives less than 40 meters (130 feet) and depth differential less than 12 meters (40 feet).

Regarding the repetitive dives, RGBM assumes microbubbles are present in the venous circulation during the surface intervals and are transferred with the blood flow to the lungs. This would reduce the surface area of the lungs, thus inhibiting offgassing. RGBM assumes that this effect will continue until the bubbles in the lungs are washed out. It calculates correction factors to cope with this, which results in more conservative decompression for repetitive dives, particularly with surface intervals of three hours or less.

Regarding multiday diving, RGBM assumes pre-existing micronuclei are excited by compression and decompression, and it takes an extended surface interval (several days) to exclude their effect. It calculates adjustments for this. According to these factors, this RGBM version adjusts the M-values downward, hence the name *reduced gradient*.

Another release of RGBM is the full iterative deep-stop version, which is more relevant to technical divers. Back in 2003-2004, Maiken from the VPM project team ran hundreds of ascent profiles through a number of commercial programs and quantitatively compared VPM-B decompressions with ascents calculated by RGBM. For a wide range of depths and gas mixtures, the schedules generated by the "full bubble" version of RGBM were highly correlated to those generated by VPM-B. For many dives, the ascent profiles calculated by the two methods were virtually identical, especially with regard to single dives. With repetitive dives, the same correction factors incorporated in the folded-over-neo-Haldanean version of RGBM come into play.

Although VPM and RGBM share the same origins, the major difference between them is that VPM does not assign a particular structure to the skin (surfactant) coating the bubbles. It just considers the skin to be either permeable or impermeable, according to the pressure exerted on the bubble. RGBM elaborates on defining the bubble skin through what Wienke calls "equation of state." This equation is central in RGBM calculations, as it controls the change in bubble size by both diffusion and Boyle's Law compensation mechanisms. VPM-0 did not account for any of that, but the enhanced VPM-B lets the bubble grow at each ascent stop according to Boyle's Law only. Bubble-diffusion mechanisms are not part of VPM-B. Another difference is that VPM makes use of information from in vitro experiments on gelatin bubbles.

RGBM is said to be based on lipid and aqueous substances. While that might hold true for the bubble-description part, it is highly unlikely for the dissolved-gas part. Just like VPM-B, RGBM incorporates a spectrum of tissue compartments, ranging from 1 minute to 720 minutes, depending on the gas mixture.[13] It is established that for a tissue with such a limited blood flow to create such a long a halftime (720 minutes, or even 600 minutes), it would not have a reasonable blood supply (i.e., oxygen) to be viable, except for teeth and the very solid portions of bones, which are not really capable of dissolving inert gas.

This arbitrary spectrum of halftimes is a model limitation that RGBM shares with other perfusion-limited decompression models. Diffusion-limited models use a different approach (to be discussed in Chapter 7). A physical reality RGBM doesn't account for is that the body tissues are interconnected not isolated, meaning that some inert gas exchange through diffusion between adjacent/adjoining tissues takes place.

Combined models
As we've seen, dual-phase (bubble) models result in much deeper stops, whereas dissolved-gas models produce longer shallow stops. Some computer manufacturers employ combined models in an attempt to get the best of both types: deep stops based on physical evidence (not just a mathematical layer like the gradient factors) and longer, shallower stops for increased safety based on models that have been in use for dozens of years. Shearwater runs the Bühlmann ZH-L16 model with gradient factors concurrently with VPM-B to produce longer, shallower stops in what they call VPM-B/GFS.

Other computer manufacturers go the other way around. They emulate a bubble model with an extra layer on top of the dissolved-gas model they use. For example, Uwatec computers incorporate the ZH-L8 ADT MB algorithm. ZH-L8 ADT is one of Bühlmann's dissolved-gas models, whereas MB is short for microbubbles — an extra layer to emulate microbubble suppression. Another example is the varying gradient model (VGM) that is incorporated in some versions of the VR3 dive computers. Yet another model is the Suunto RGBM, which is a bubble layer folded over a spectrum of M-values to reduce the permitted supersaturation gradients.

The major drawback of this combined-model approach is that it works fine in certain regions only. When the dive gets out of these regions (longer and/or deeper), it tends to produce unrealistically long schedules.

Summary
In this chapter, we clarified that bubbles form after almost every dive, even if the diver is not experiencing any problems. This bubble formation has been attributed to the presence of preformed "seeds" capable of growing during the ascent phase. We then presented an overview of some important concepts related to the relatively new approach of staging

divers via controlling both the individual bubble size and the total bubble volume. The evolution of these ideas were explained. Several algorithms were demonstrated and special attention was given to the most popular one (VPM-B). Model variations were discussed, illustrating the idea behind each of them. Finally, the concept of combining models was highlighted and the advantages and drawbacks of this approach were detailed.

CHAPTER 7
Other Decompression Models

Slab diffusion

In 1952 Dr. Henry Valance Hempleman (1922-2006) introduced a new theory of decompression. His new approach was more than just a modification of Haldane's work. Hempleman noticed the majority of DCS symptoms were manifested as limb pain, in or around the joints. He suggested that a single tissue was responsible, so he based his model on one tissue only. However, a single tissue couldn't fit the range of decompression if a perfusion-limited model was used. So Hempleman assumed *diffusion* rather than perfusion would be how inert gas was absorbed in his single, critical tissue.[1] His assumption appeared to be consistent with the poor perfusion in the cartilages around the joints.

Hempleman modeled the joint as a slab of tissue lacking blood vessels. One face of the slab was well perfused, and the inert gas was diffused from this face to the inner parts of the slab. The inert gas supply to the slab was defined by a diffusion equation rather than compartments and halftimes. After some mathematical derivations, Hempleman concluded the quantity of inert gas entering the slab would be equal to the ambient pressure multiplied by the square root of time.

$$Q = \text{Ambient Pressure} * \text{SQRT (Time)}$$

The NDLs produced by Hempleman's model matched to a great extent those of the U.S. Navy (empirically derived). For decompression dives, the model staged the diver according to the same equation, along with an arbitrarily chosen tolerable level constant.

In 1960 Hempleman challenged another basic assumption of Haldane's, which was gas uptake and elimination took identical times. He presented evidence from animal studies that the uptake and elimination of inert gases were not symmetrical and assumed the elimination process was one and a half times slower than that of the uptake.[2] This is what we now refer to as asymmetric gas kinetics. The Royal Navy adopted Hempleman's work. Known as Royal Navy Physiological Laboratory (RNPL) tables, they were first published in 1968; a modified version was published in 1972. The BSAC adopted a version of the 1972 modified tables until they published their own set in 1988.

Kidd–Stubbs

In 1962 Derek Kidd and Roy Stubbs started a program to develop a real-time dive instrument to calculate the required decompression obligation according to the depth-time history. Initial versions of this instrument were pneumatic, mechanical, analog computers, frequently referred to as Kidd–Stubbs pneumatic analog dive computers (K-S PADC).

Cavities into which the gas could enter at a controlled rate and pneumatic resistors consisting of micropores were used to simulate the different tissues in the human body and to describe the gas flow in and out of them. At first, Kidd and Stubbs used four compartments connected in parallel, which means they do not affect each other, along with the U.S.

Navy algorithm to generate the required decompression schedule according to the real-time readings. They made a series of modifications to halftimes and supersaturation ratios to accommodate a wide spectrum of single and repetitive dives. The underlying mathematical expressions described the operation of a piece of hardware aiming to generate safe decompression schedules rather than a physiological model for decompression.

Kidd and Stubbs believed the gas transfer between the lungs and the body tissues would be better described by an interconnected series of compartments. Their assumption is an extension of the slab-diffusion concept. They arranged the four compartments in series, and assumed they are all bearing risk of DCS.

In 1970 it was discovered that an inherent distrust in the safety of the model for deep dives led the hyperbaric chamber operators to stay deeper than the computed safe ascent depth (SAD) by as much as 3 meters (10 feet). When the operators were instructed to follow the model exactly, the result was a whopping 20% incidence of DCS, compared with only 3.6% when they applied their own adjustments. The affected range was 60-91.5 meters (200-300 feet). In 1971 Stubbs analyzed the results and applied some correction factors, publishing the model later the same year.[3] The Defence and Civil Institute of Environmental Medicine (DCIEM) of Canada adopted it as a safer alternative to the U.S. Navy tables.

During the 1970s many single and repetitive dives were conducted using the pneumatic computer incorporating the 1971 algorithm. A microprocessor version of the dive computer was developed in the late 1970s.

In 1979 a critical study was initiated to evaluate the performance on the dive computer. The algorithm was generally safe, yet it was overconservative in some regions and overaggressive in others. In 1983 a modified version called DCIEM 1983 was derived. The new derivative had risk of DCS associated with only the two outermost compartments of the series. The two innermost compartments were nonrisk-bearing yet they influenced the risk of DCS indirectly.[4] In 1992 the DCIEM manual was published with new tables generated by the modified model. Although generally less conservative than the former 1971 version, this set of tables enjoyed wide acceptance by the diving community and was considered one of the safest and most conservative. The modified model was also incorporated in one of Citizen's line of watches.

U.S. Navy exponential linear (USN E-L)

Haldane presumed that if no symptoms of DCS were present postdecompression, then no bubbles were formed in the blood. He also assumed that the uptake and elimination of inert gas by the tissue compartments were symmetrical and they took an exponential form.

When Doppler tests were incorporated into decompression research, they illustrated that bubbles could actually form in the blood even if no symptoms of DCS were present postdecompression. The presence of these bubbles would affect the offgassing efficiency, meaning it would slow the offgassing rate, which in turn would cause more bubbles to form.

To compensate for the anticipated decrease in offgassing rates caused by the presence of these bubbles, Dr. Edward Thalmann (1945-2004) published in 1984 an algorithm based on the asymmetric gas kinetics concept, meaning the uptake and elimination of inert gas by the tissue compartments were *not* symmetrical.[5] Called the USN E-L (also known as VVal-18 Thalmann) algorithm, it assumed the inert gas was absorbed by tissues at an **exponential** rate (as in other neo-Haldanean models) but was discharged at a slower **linear** rate (hence the name E-L). This not only resulted in longer decompression stops but also affected the offgassing rate during the surface interval, causing it to slow down. In conclusion, the schedule generated for the first dive would be more conservative, and the schedules generated for repetitive dives would be much more conservative.

In 2001 some U.S. Navy divers made the first official computerized decompression dives in U.S. military history, using dive computers made by Cochran Undersea Technology and incorporating the USN E-L algorithm. In 2012 the Navy dive computers were validated by faithful replication of "gold standard" software implementations of the E-L algorithm maintained by the U.S. Navy.[6]

Continuous research by the Navy Experimental Diving Unit (NEDU) on the E-L algorithm resulted in a set of tables based on Thalmann's work and produced by Gerth and Doolette in 2007.[7]

Probabilistic models

A probabilistic decompression model is a statistical inference model validated with enough manned dives with known outcomes (usually several hundred to several thousand). The model is first developed then fitted to a dataset containing as many human dive profiles as possible, as long as the incidence of DCS in these profiles is precisely known. The dataset should accurately describe the type of dives the model is equipped to handle (air-only, oxygen-nitrogen, oxygen-helium, oxygen-nitrogen-helium, high O_2 content, etc.).

In the mid-1990s the U.S. Navy had a dataset consisting of 3,222 oxygen-nitrogen dives. The probabilistic models calibrated with this dataset were successful in describing DCS risk observed across a wide variety of oxygen-nitrogen dives yet failed to account for the observed DCS incidence in dives with high ppO_2 during decompression. The U.S. Navy researchers added 1,013 O_2 decompression dives to the calibration data and refitted the models to the expanded dataset.[8] The efficiency of predicting the observed incidence in dives with pure oxygen breathing during decompression increased from 40% to 90%.

Probabilistic decompression models do not try to predict the incidence of DCS following a specific dive profile because this falls into the "statistically uncertain" category. Instead they generate decompression schedules to any level of accepted risk. Divers can then compare these generated schedules with the dive plan to calculate the probability of DCS occurrence.

Thalmann and his colleagues conducted the first validation trial of a probabilistic decompression model from 1991 to 1992 by incorporating the exponential-linear kinetics.[9] Other decompression models followed, some considered risk as a function of tissue compartment gas load whereas others considered risk as a function of bubble volume. The latter hypothesized that symptoms of DCS appeared when the total volume of bubbles in a unit volume of any tissue exceeded the critical specific volume of a free gas phase.

A study was conducted in 2010 to determine which decompression gas was less risky with respect to DCS.[10] Using a commercial software package, both VPM-B and Bühlmann with gradient factors tables were produced for EAN32 dives with either EAN50 or pure O_2 for decompression. Three probabilistic inference models were used to generate probability of DCS-based risk profiles for the dive tables.

Although the study did not mention the total run times, conservatism levels, ascent and descent rates, Bühlmann's model (probably ZH-L16B, but ZH-L16C wouldn't be an unusual choice) or version of the commercial software package (default software settings were used except to adjust the depth of the last decompression stop to 6 meters [20 feet]), it concluded that, irrespective of the overall decompression time (which we didn't know anyway), using oxygen held less probability of DCS compared with EAN50. The results indicated that the probability of DCS risk was always lower when Bühlmann with gradient factors tables were in use, but we couldn't count on that either, as we don't know the VPM-B conservatism level, Bühlmann's model or the gradient factors the authors used.

In 2004 a probabilistic model was developed that assumed a random pattern of bubbling processes in living tissues. The developer of this model claimed that it explained why resistance to DCS in mammals increased with a lower body mass and greater specific blood flow in tissues.[11]

Developing a probabilistic inference model is not a difficult task for computer scientists, especially those in the artificial intelligence field. To make it work properly, however, the model has to be "sufficiently trained." This training requires a dataset representative to the modeled phenomenon, which is not available off-shelf. The dataset should match the dive plans for which the diver wants to assess the level of risk. For example, if the diver used a probabilistic model fitted to an air-only dataset to assess a trimix dive, the result would be comparable to using an English-only OCR system to try to recognize Arabic images.

So unlike other decompression-planning tools, the diver can't just buy a probabilistic model. It is important not to confuse probabilistic models with dive-profile analyzers. A profile analyzer is usually an algorithm that compares the dive profile entered by the diver with a very conservative schedule it generates on the fly. Looking at the two plans concurrently, if the entered profile has shorter time frames, the corresponding parts are identified as potential risk of DCS. If an ICD formula is incorporated, the dive-profile analyzer can also display ICD warnings on gas-switch stops.

Decompression Computation and Analysis Program (DCAP)

In the late 1960s physiologist Dr. Robert (Bill) Hamilton (1930-2011) and engineer and programmer David Kenyon worked in a laboratory dedicated to commercial diving. Their laboratory was charged with the task of developing tables for diving to a maximum depth of 200 meters (660 feet). The laboratory's efforts were successful. Afterward, Schreiner, the laboratory's chief decompression expert, moved to a management position, and both Hamilton and Kenyon eventually acted as independent decompression consultants.

Kenyon refined Schreiner's concepts into a decompression computation program. Hamilton and Kenyon used this program to furnish specially developed tables to navies and commercial diving companies. However, due to the lack of business communication tools such as faxes and emails at that time, they had trouble providing support to overseas clients. One of their customers, the Swedish Navy, proposed to install the table-generation software on their computer so that they could make whatever tables they would need without communicating with the decompression expert house. The table-generation software, called DCAP, incorporated several underlying decompression models. Afterward, DCAP was acquired by several key accounts, including navies, commercial diving firms and research laboratories. Scores of special sets of tables generated for specific projects were acquired by major customers, one of which was the NOAA-Hamilton Trimix Decompression Tables used by NOAA for diving on the American Civil War vessel USS *Monitor*.

Although DCAP uses various computational models, it is most famous as the Tonawanda IIa wrapper. Tonawanda IIa (frequently referred to as Hamilton-Kenyon) is described by its developers as a Haldane-Workman-Schreiner algorithm that accommodates multiple mixtures and can employ up to 20 compartments. It allows the user to control both the halftimes and the M-values. A well-known subset of Tonawanda IIa is the 11F6, which uses 11 halftime compartments ranging from 5 to 670 minutes (for nitrogen) along with the corresponding M-values. This matrix of M-values is referred to as either MM11F6 (when meters are used) or MF11F6 (when feet are used). Other models handled by DCAP include Kidd Stubbs, VPM and USN E L. Tonawanda IIa has also been folded over VPM to generate more conservative schedules.

The major advantage of DCAP is that it can apply constraints to decompression algorithms to guide the ascent. One of these constraints is a model called t-delta-P, which is folded over the algorithms DCAP incorporates. It uses the integral of supersaturation over time as an additional ascent limit. In other words, it adjusts the decompression to account for time spent under supersaturation, and it frequently recommends recalculating the ascent schedule at a slower rate. DCAP also affords a statistical estimate of the reliability of the schedules it generates, meaning that it can estimate the probability of occurrence of DCS of its generated schedules. For saturation (commercial) diving, DCAP employs a function that links the ascent rate to the oxygen level. Other capabilities include generating schedules for deep dives with neon and forecasting ICD.

Hamilton Research Ltd., the owner of DCAP, did not encourage small diving contractors to acquire or operate DCAP, mainly because of its cost and the necessary training and experience. 11F6 with its matrix of M-values (M11F6 algorithm) is now implemented in Ultimate Planner.

Arterial bubble (AB)

"If small bubbles are carried through the lung capillaries and pass, for instance, to a slowly desaturating part of the spinal cord, they will there increase in size and may produce serious blockage of the circulation or direct mechanical damage."
— *J.S. Haldane. The prevention of compressed air illness. J. Hygiene Camb. 1908 p. 352*

The AB model assumes that the filtering capacity of the lungs is a critical issue. As a bubble filter, the lungs get rid of the bubbles on the venous side through ventilation. However, smaller bubbles are suspected to pass to the arterial side of the circulation. A study using a contrast agent called Levovist was conducted in 2002,[12] confirming this assumption. Levovist contained stable gas bubbles with calibrated sizes between 3 and 8 micron trapped in galactose. Levovist was then injected intravenously, and measurements were made a few minutes later in the cerebral, renal and lower limb arteries. Levovist bubbles were found in the arterial side of the circulation, which indicated that decompression bubbles could also travel through the lungs and cross to the arterial side, especially at the beginning of the ascent where the bubbles were still small. The threshold radius was suspected to be the size of a blood cell. During the ascent, the bubbles grew, so the lungs trapped them more efficiently.

Just like both the dissolved-gas and the dual-phase models, AB presumes that when the ascent starts, the compartments start to offgas as soon as a gradient is created. The bubbles are then transported through the venous system to the lungs, which in turn filter the bubbles by trapping them in the capillaries according to their respective diameters then eliminating them through the alveoli. However, small bubbles may cross the lungs, passing to the arterial side of the circulation. The distribution of the blood at the aorta is such that a crossing bubble is likely to reach a neurological location such as the brain or the spinal cord; both are fast, well-perfused tissues, and both have the capability of acting as gas reservoirs and diffuse gas into the crossing bubble. The crossing bubble would either proceed to the venous side or grow onsite, causing serious neurological DCS, not just limb pain or skin rash.

The main advantage of AB is that trying to simulate the efficiency of the lungs and associating it with the DCS susceptibility responds to many questions with unquantified answers. A better diver is a better bubble filter, one with more efficient lungs. Regarding the individual susceptibility, simulating the efficiency of the lungs would account not only for the interindividual variability (age, obesity, smoking, etc.) but also for the intraindividual variability (fatigue, dehydration, etc.). Regarding the CO_2 production during the dive, it is reasonable to assume that excess CO_2 would reduce the filtering capacity of the lungs and thus would cause bubbles to pass to the arterial system. So diving situations associated with the production of excess CO_2 (anxiety, stress, temperature variation, exhaustion, hyperventilation, high breathing resistance, etc.) would be automatically accounted for.

Another advantage of the AB is that trying to simulate the efficiency of the lungs and associating it with the DCS susceptibility is consistent with the accidental production of arterial bubbles found in several cases such as patent foramen ovale (PFO, to be discussed in Chapter 8), for example. It also offers an explanation for the occurrence of neurological DCS (also known as CNS DCS).

A proposed scenario is that when the ascent commences, some bubbles are trapped in the lungs waiting to be filtered out with the exhalation. If the diver descends again, these bubbles will experience size reduction due to the excess pressure exerted on them. Some would pass through the lung capillaries and eventually dump into the arterial bed. This might explain why divers should avoid multilevel recompression (sawtooth) profiles. This hypothesis also suggests a probable explanation of the "undeserved" hits. Multilevel dives will be covered in Chapter 8.

AB provides a possible explanation for the criticality of the initial ascent phase. Symptomatic bubbles are not necessarily generated onsite. AB assumes the process of small bubbles passing to the arterial system during the initial ascent phase is an early prediction of possible DCS symptoms. Once the bubbles reach a critical size, they are either filtered in the lungs or stopped at the tissue level.

AB also provides another interesting presumption. When the last decompression stop is at 3 meters (10 feet), the slight variation in depth due to less-than-perfect buoyancy skills or rough seas would produce sufficient pressure changes to reduce the bubble size, facilitating its escape to the arterial bed. That is why the AB model advocates performing the last stop at 6 meters (20 feet), as the same variation in depth will result in less change in pressure.

Hills had the same observation in the early 1960s when he was investigating the decompression procedures followed by the pearl divers. They used to drop the last stop(s) and ascend directly to the surface from 6-9 meters (20-30 feet). An early attempt was made to turn this scenario into a decompression model. The resulting model (now called AB-1) developed by Jean Pierre Imbert incorporated the concept of compartment halftimes. For the safe ascent criteria, the formulation of arterial bubbles was suspected but not established in mathematical terms. Instead, an approximation was defined empirically by fitting mathematical expressions to selected exposures stored in the database of the French commercial diving giant Comex. AB-1 was used to compute a set of decompression tables, which were sent offshore for evaluation on selected Comex worksites. In 1986, after some minor adjustments, the tables were included in Comex diving manuals and were used as standard procedures. In 1992 the tables were included in the French diving regulations under the name of Tables du Ministère du Travail 1992, or simply MT92 tables.

In the 1990s Imbert introduced a second version called AB-2, a dual-phase model. It amended the equation that describes the ascent criteria by adding the arterial inert gas tension as a parameter. It also incorporated the concept of varying compartment halftimes. The result was a deeper stop than neo-Haldanean models.[13] The generated schedules were comparable to Bühlmann's ZH-L16 with either gradient factors or RGBM folded over. Just like other models, AB-2 did not specify the ascent rate. It was a user input. For this parameter, AB-2 would expect an input between 6 and 9 meters/min (20 and 30 feet/min).

AB-2 was first calibrated by data fitting on both Comex database and experimental dives using air, nitrox and heliox. AB-2 was later recalibrated using experimental bounce dives conducted by the French Navy. Since 1999 it has been used by some sports divers in Tunisia and Corsica.

Copernicus

A study was initiated in 2003 to determine the effect of decompression schedules on bubble formation following surface decompression using oxygen.[14] A pig was compressed to 40 meters (130 feet) for 90 minutes while breathing air. Three different ascent profiles were tested. The first was a USN staged decompression profile. The second was a profile using linear continuous decompression with the same total decompression time as the USN profile. The third was a linear profile with half the total decompression time compared with the first two profiles. The subsequent surface decompression at 12 meters (40 feet) lasted 68 minutes for all three schedules. The study illustrated that following the final decompression, the two linear schedules produced the lowest amount of vascular gas, with the fastest schedule producing significantly less bubbles in both the pulmonary artery and the jugular vein than the other two. This was an experimental demonstration that a significant reduction in decompression time could be achieved with a dramatic reduction in bubble formation.

So the key word is optimization. All decompression algorithms calculate the time to stay at each depth based on either supersaturation or bubble suppression. However, current models don't consider the initial ascent phase, although both experimental evidence and bubble growth theory suggest that bubble growth starts at that phase.

Another important aspect is that a lot of decompression models are empirically derived. For example, while developing ZH-L, Bühlmann's endpoint was the clinical symptoms of DCS. This approach requires an extensive amount of empirical data yet does not validate the actual gas dynamics and mechanisms behind DCS. The result is that the models are usually safe enough only in the regions they were developed, thus extrapolating them (mathematically deriving a model for helium from the existing nitrogen data, for example) would lead to less safe consequences.

Copernicus presumes that the evolution of vascular bubbles is strongly linked to the risk of serious DCS. Just like other dual-phase models, it tries to describe both the dissolved gas tensions and the distribution and growth of gas bubbles in the diver's body. The new feature is that it also tries to predict the mechanism for injury by the bubbles. In other words, the endpoint for Copernicus is not the clinical symptoms of DCS, it is the bubble formation in the diver's body. Using this approach, the development team led by Dr. Alf Brubakk will not need the extensive amount of empirical data Bühlmann needed, yet the validation will be more reliable.

Copernicus consists of a descriptive model of the mechanisms behind the occurrence of serious DCS (based on the evolution of vascular bubbles), a dynamic optimization algorithm to control the model parameters and a validation strategy through bubble measurement rather than the conventional DCS/no-DCS endpoint.

The model inputs are ambient pressure, breathing-gas composition and blood perfusion. During decompression, Copernicus will manipulate the ambient pressure (depth) and the gas composition to achieve the wanted outcome. Blood perfusion is estimated through measurements and is not an input the model wants to control. The output is a description of the evolved bubble spectra in the body. The optimization phase looks at getting the diver as fast as possible to the surface without letting the decompression stress exceed an accepted bubble score.[15]

Copernicus is still under development. One of its tentative conclusions is that dives with long bottom time (30 minutes or more) may benefit from deep stops, whereas dives with bottom times less than 30 minutes will not benefit from them.

SAUL

In 2007 Dr. Saul Goldman proposed a new model based on the interconnectivity of body tissues.[16] The concept behind his model is that dissolved materials (drugs, for example) could diffuse from one tissue to another. This is an established concept, and our current understanding is that it holds true for inert gases like nitrogen and helium.

What is new with this model? The Kidd-Stubbs model is already diffusion-limited and involves four compartments connected in series. The SAUL "general" model presumes three compartments *interconnected in parallel*. The difference is that unlike Kidd-Stubbs, all compartments in the SAUL model are connected to the blood flow — in parallel. However, only the central compartment is well perfused. Unlike perfusion-limited models, the peripheral compartments are connected to the central, well-perfused one — in series. The risk of DCS is carried entirely by this central, well-perfused compartment, while the peripheral compartments are not, in themselves, risk bearing. They only influence the risk of DCS indirectly by acting as reservoirs of dissolved inert gas.

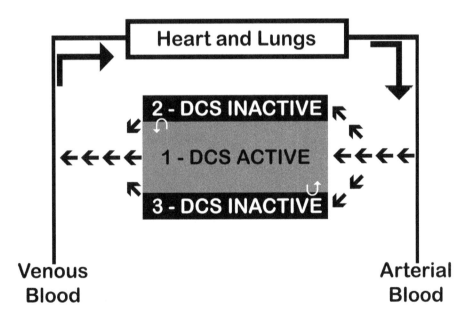

SAUL interconnected model

When the body is ongassing, the peripheral compartments act as overflow tanks. When the offgassing phase starts, the sink becomes a source and the "overflow" gas diffuses back into the risk-bearing compartment. For example, a fatty tissue is not bearing the risk of DCS in itself. However, it can supply one of the nerve or the connective tissues next to it with dissolved inert gas via diffusion during the ascent, this increasing the receiving tissue's risk of DCS. On the other hand, in the Kidd-Stubbs original model all four compartments bear risk of DCS. The new derivative has risk associated with only the two outermost compartments of the series. Goldman claims that when an explicit risk is put into more than one compartment, the prediction capability of the model deteriorates.

The SAUL model is mainly about gas-exchange dynamics. Currently, two versions of SAUL exist: SAUL ICM (interconnected model) and SAUL ICBM (interconnected bubble model). Adding microbubbles to the model produces minor improvements. The most intriguing question about this model is does SAUL really stand for Safe Advanced Underwater aLgorithm, or is it simply the modeler's first name?

Probability and severity

In 2017, a study investigating a new method to predict the probability of DCS was published. Case reports for 3322 air and nitrox dives resulting in 190 DCS events were retrospectively analyzed and the outcomes were scored as the following:

1. serious neurological

2. cardiopulmonary

3. mild neurological

4. pain

5. lymphatic or skin

6. constitutional or nonspecific manifestations.

Following standard U.S. Navy medical definitions, the data were grouped into mild (Type I: manifestations 4–6) and serious (Type II: manifestations 1–3). The researchers also considered an alternative grouping of mild (Type A: manifestations 3–6) and serious (Type B: manifestations 1 and 2). The current U.S. Navy guidance allows for a 2% probability of mild DCS and a 0.1% probability of serious DCS. Alternatively, the researchers developed a hierarchical trinomial (3-state) probabilistic model that simultaneously predicts the probability of mild and serious DCS given a dive exposure.[17] Both the Type I/II and Type A/B discriminations of mild and serious DCS resulted in a highly significant improvement in trinomial model fit over the binomial (2-state) model. According to the presented results, it would be better to replace the currently used DCS / No-DCS model and consider a new one employing Serious-DCS / Mild-DCS / No-DCS.

Instead of planning your dive the way you do right now, you would enter your accepted risks of serious and mild DCS. The lower your accepted risks are, the longer the generated schedule would be.

Summary

In chapter 3 and 6, we presented some of the most popular dissolved-gas and dual-phase decompression models. In this chapter, we introduced other models based on different assumptions.

In the 1950s, Hempleman assumed diffusion rather than perfusion would be how inert gas was absorbed. Accordingly, he proposed a single critical tissue model. A decade later, Kidd and Stubbs extended Hempleman's concept, proposing a model based on four compartments arranged in series.

In the 1980s, Thalmann published an algorithm based on the asymmetric gas kinetics concept. A decade later, the employment of statistical inference models in the field of decompression started.

Another approach was followed by Hamilton and Kenyon. They applied "constraints" to various decompression algorithms in order to guide the ascent. Throughout the 1980s, Imbert worked on a model assuming that the filtering capacity of the lungs was critical and smaller bubbles were suspected to pass to the arterial side of the circulation. More recent models focused on issues like predicting the mechanism for injury by the bubbles, predicting the probability of DCS, and the interconnectivity of body tissues.

CHAPTER 8
Various Topics

The oxygen window

The term "oxygen window" was first coined by Albert Behnke in 1967. The same phenomenon is frequently referred to as *inherent unsaturation* or *partial pressure vacancy*. It is defined as the difference between the ppO_2 in the arterial blood and that in the venous blood. This ppO_2 drop is caused by metabolism and is believed to reduce the chances of bubble formation.[1]

Under normal conditions (breathing air at 1 ATA), most O_2 is transported in the blood bound to hemoglobin. When O_2 is bound to hemoglobin, it is no longer dissolved in solution and no longer contributes to the ppO_2. The binding of O_2 by hemoglobin causes a ppO_2 difference between the tissues and the arterial blood. When blood perfusion takes place, O_2 moves into the tissues and CO_2 moves into the blood. On a mole-for-mole basis, the production of CO_2 is almost the same as the consumption of O_2. If the ppO_2 absorbed from the blood were replaced by an equal $ppCO_2$ from the tissues, there would be no change in the total partial pressure between the arterial and the venous sides. This latter condition does not get fulfilled. As the blood traverses the tissues, the increase in $ppCO_2$ is much less than the decrease in ppO_2, as not all of the consumed O_2 gets converted to CO_2. Under normal conditions, about 20% of the consumed O_2 doesn't convert.[2] Also the solubility of CO_2 in the blood is some 20 times higher than that of the O_2. We already know that higher solubility means more gas dissolved, which in turn means that a tissue (or blood) will absorb a larger volume of a highly soluble gas as opposed to a lower-solubility gas before reaching any given partial pressure. So if a given volume of gas dissolves in a particular tissue, the tissue loading of a highly soluble gas will be lower than the tissue loading of a low-solubility gas. Applying this to CO_2 and O_2, the higher solubility of CO_2 means that it will produce less partial pressure than O_2 would have produced. In conclusion, the ppO_2 decrease is estimated to be 10 times the $ppCO_2$ increase.

Levels of O_2 and CO_2 in the tissues can influence the blood flow. The oxygen window provides a tendency for absorption of the gas quantities in the body such as DCS bubbles. With DCS bubbles, the oxygen window is a major factor in the rate of bubble shrinkage when the diver is in a steady state (no inert gas uptake or elimination taking place). It also modifies bubble dynamics when inert gas is being taken up or eliminated by the tissues and may sometimes prevent the transformation of bubble nuclei into stable bubbles.[3]

However, reducing the chances of bubble formation does not mean better inert gas elimination. Powell clarifies: *"Many believe that numerous gas bubbles in the pulmonary blood vessels can modify gas exchange, limiting it, in fact. We looked into this and took advantage of the fact that altitude depress often produces copious numbers of bubbles in a high percent of the test subjects. Techniques are available that allow for the measurement of what is termed 'pulmonary dead space.' This is what occurs in the lungs when air can enter and exit the alveoli (little air sacks) but blood does not flow to them because the capillaries are blocked. In our case, the blockade would be caused by gas bubbles. The test is simple and noninvasive and requires the ability to measure carbon dioxide in the arterial blood and the exhaled breath. To our mild surprise, no evidence of impairment of gas exchange was found. This means that bubbles do not modify nitrogen (or any other inert gas) elimination even when a great many gas bubbles are present in the lungs. Individuals exposed to our high altitudes produced more bubbles than you would ever expect in a recreational diver.*" [4]

By switching to a richer mix during decompression, it is believed the oxygen window gets "wider" as more oxygen is introduced to the body. This is often referred to as "opening the oxygen window." However, for this opening to have the desired effect of reducing the chances of bubble formation, it is believed by some that regardless of what your favorite decompression-planning tool dictates, you should extend the gas-switch stop so that blood circulates around the entire body. The common practice is to extend the gas-switch stops by at least three minutes. To project the effect of extending the gas-switch stops on the decompression schedule, Ultimate Planner has an "Extended Stops" feature that will account for any extra minutes spent on the gas-switch stops and alter the decompression plan accordingly.

Flying after diving (FAD)

At sea level, the pressure exerted on our bodies by the column of air above us is 1 atmospheric pressure — approximately 1.01325 bar at 20°C (68°F). As we go above sea level, this pressure decreases as the column of air above us gets shorter. The temperature also decreases by approximately 1°C (1.8°F) every 150 meters (492 feet). This decrease in temperature affects the air density (and accordingly the pressure it exerts) in an inversely proportional manner.

As volume is inversely proportional with pressure, when we ascend to altitude or fly after diving, the size of inert gas bubbles in our bodies tends to increase. To avoid getting a DCS hit, we have to wait for some time before flying or ascending to altitude.

To determine the required preflight surface interval, DAN conducted an FAD workshop in May 2002. The consensus guidelines that were established are as follows: [5]

- For a single no-stop dive, a minimum preflight surface interval of 12 hours is suggested.

- For multiple dives per day or multiple days of diving, a minimum preflight surface interval of 18 hours is suggested.

- For dives requiring decompression stops, there is little experimental or published evidence on which to base a recommendation. A preflight surface interval substantially longer than 18 hours appears prudent.

All recommendations assume air dives followed by flights at cabin altitudes of 2,000 to 8,000 feet (610 to 2,438 meters) for divers who do not have symptoms of DCS. The recommended preflight surface intervals do not guarantee avoidance of DCS.

Dives that contributed to these guidelines were simulated with subjects remaining dry and at rest. The risk of immersed and exercising dives may be greater. For increased safety, dive operators usually advise a minimum preflight surface interval of 24 hours for both recreational and technical dives. Several dive computer brands recommend the same, whereas others do attempt to calculate the minimum preflight surface interval. They simply consider that the surface is "shifted up" to the cabin altitude and calculate an obligatory stop time (on air) accordingly. Some manufacturers assume double the maximum cabin altitude (16,000 feet or 4,876 meters) when trying to calculate the preflight surface interval, as when the correct numbers are used, the resulting no-fly times are much shorter than DAN's recommendations.

Accelerating no-fly time

In 2010 I considered the possibility of using surface oxygen to accelerate the suggested preflight surface interval. At the end of the day, this surface interval is nothing but a mandatory stop on air for desaturation. Will using pure oxygen instead of air reduce this surface interval? And if it will, how can we calculate the time gain? I consulted a group of the world's most accomplished divers to collect some firsthand information. Their comments are as follows:[6]

Bret Gilliam, founder of TDI, started diving in 1958. His experience includes military and commercial projects.

"We began using protocols of surface O_2 breathing to reduce time to fly as far back as the early 1970s — very simple procedures. We started to use surface oxygen breathing to allow us to get on planes faster to get out of some areas the same day that we did some diving in the morning. Our protocol was to cut four hours off the no-fly time for every one hour on surface 100% O_2 via mask. It's pretty simple when you think about it. Normal atmospheric air is 21% oxygen. Increase your breathing intake to 100% O_2, and you gain a 400% advantage. Plus it accelerates inert gas removal even beyond that initially and then begins to taper off exponentially. But we could easily do a couple of long dives before noon, breathe O_2 for an hour and fly out later that evening. No problems ever.

"This all began back in the days of dive tables when the basic Haldanean model that the U.S. Navy used was standard practice. There were no personal computers and no dive computers. All dive profiles were 'square' except for those of us who used the SOS/ Scubapro Decompression Meter. (By the way, those devices worked fine for no-deco multilevel diving. Just never let it go into deco, or you could be hanging forever.)

"Today, virtually all decompression algorithms incorporated into dive computers are far more conservative than the Navy model, so the surface O_2 breathing protocols are probably even more applicable.

"It's all about the priority of your travel and what level of risk tolerance you are willing to accept. I have never subscribed to the '24-hour sit-down rule' before flying. It just doesn't make sense, and most informed professionals chose other rules a long time ago.

"There are literally scores of algorithms that can be applied to a mathematical model to expedite inert gas outflow at the surface. It both relieves any subclinical decompression stress and allows shorter intervals before flying. Remember that most

commercial aircraft have cabin pressures of about 8,000 feet (2,438 meters). That's not a huge differential.

"What works for me and my diving partners may not work for all. But I'm 59 and have been doing this now for nearly four decades professionally. I've never had an incident."

Steve Lewis is an instructor trainer who served as a member of the TDI Training Advisory Panel and as director of product development.

"I am aware of empirical data — been doing it for almost 20 years. The most extreme case was a three-hour dive in Eagle's Nest with 30 minutes of surface oxygen immediately postdive (about 19:45), 15 minutes in the morning and on a flight at 08:00. Not a niggle.

"Several years ago we had only rudimentary tables and little hard data and no customizable decompression software. Everything we did was seat of the pants. We would work off what was then the general guideline of 12 hours no-fly time, and we would surface breathe pure oxygen via a demand mask. For every minute spent on the mask, we would assume three to four minutes of no-fly time had evaporated. Essentially, an hour on the mask immediately postdive meant that our no-fly time was now eight hours (three-hour acceleration) or seven hours (four-hour acceleration). We rarely spent more than an hour on the mask at a time because it is so uncomfortable. However, on a couple of occasions I wore a mask for up to three hours following deep trimix dives and flew immediately. Well there was the screwing around at the airport for an hour or so, but you get the picture. We never had any issues with niggles or bends.

"I do not know of any incidents of divers flying higher than 6,500 feet (1,980 meters) in unpressurized aircrafts postdive and do not have any figures or stories to tell of people getting bent flying within 12 hours when they have used 'oxygen saturation therapy.'

"These days I am more circumspect regarding NOAA 24-hour CNS limits, but I still use the old numbers 3:1/4:1 ratio. In addition, I spend longer than VPM calls for at 4 meters (13 feet) breathing pure oxygen or an 85/15 heliox if a flight is in my immediate future. For example, if the table calls for a total of 35 minutes decompressing at 6 and 3 meters (20 and 10 feet), I will do the required 6-meter (20-foot) stop and then spend 1.5 times the total time at 4 or 5 meters (13 or 16 feet), notwithstanding my CNS status, which always governs my behavior.

"With regard to your question about the tables for timing surface intervals, not sure how you would work it, but I would be interested in the results."

Mark Powell, an instructor since 1994, is an accomplished author. His favorite topic is decompression theory.

"I have used O_2 to extend deco stops and on the surface to reduce the risk at the end of trips. I would generally add 5 to 10 minutes onto last stop and then surface breathe for up to 30 minutes after the last dive. However, I have never measured or calculated this; it is more of an ad-hoc practice."

Ben Reymenants, a training director with TDI, has helped write some of their manuals.

"I don't use surface O_2 to accelerate no-fly time. But I use it when I have an extremely heavy dive or when I suffer subclinical decompression illness. You could use it to allow divers that have done heavy multiple-day diving to jump on a plane within 24 hours. This way you give them peace of mind. Keep into consideration that divers jump on the plane tired and dehydrated and then start drinking alcohol, which increases the risk.

"Even using complex decompression software will not allow you to clean up tissues to a level that critical bubble radius is sufficiently reduced allowing for hypobaric exposure. In the old days you could fetch a ride in the chamber on O_2 and then step on a plane in 12 hours.

"Take into consideration that there are other enhancers — such as rest, surface temperature, personal metabolism and hydration — where software has no grip on. I don't see a place for this train of thought. Be well aware that we are talking slow tissues offgassing here. Surface O_2 will reduce critical bubble tension in fast tissues but will also induce peripheral vasoconstriction. At this point it is unknown if this can actually slow down the offgassing of the slowest tissues.

"The software can only hope for those enhancers, and also this exposure is short term, whereas at the surface it is long term. Is the person resting? Running around? Sleeping?

"I used to accelerate decompression in chambers on pure oxygen, at the surface and at 20-meter (66-foot) depth. A few times I had transient skin bends just because I was cold during deco."

Brett Hemphill, a cave diver since 1990, is a technical consultant working on protocols and equipment configurations for safer cave diving with CCRs.

"There is some merit to the question you are asking concerning possibly expediting no-fly times after being in decompression.

"I am sure padding decompression, surface O_2 and additional hydration might possibly decrease the onset of decompression sickness brought on by flight in a noncompressed plane, but using any collaborative data to promote this venture probably would not be accepted widely.

"Another good example would be a dive my friends and I will be doing in a week or two — several 90-meter (300-foot) dives within a week. We will continue to increase our bottom time over every other day for three days. With residual nitrogen being present, how can we keep from elongating upcoming decompressions without incurring DCS?

"This is a somewhat relative comparison to the question you are asking. I will be increasing my 12-meter (40-foot) stop on every subsequent dive to optimize the offgassing of denser tissue compartments that just don't respond as effectively at 6 meters (20 feet) as they would at 12 meters (40 feet). I will not essentially be running higher O_2 at this depth but certainly maintain a smooth curve while monitoring my

---

O_2 clock, as I believe can be turned back and reset, especially at this depth where ongassing just doesn't occur.

"Ultimately if a person must fly and cannot perform an adequate amount of additional decompression, I am sure that additional O_2 and hydration would certainly lessen the possibility of altitude-type DCS, but even Dr. Bruce Wienke, who at this time I still consider the most knowledgeable dive gas physicist, would never publish or promote any more than what I've possibly suggested. Any possible gains one might achieve couldn't outweigh the equally possible adverse affects."

In conclusion, we have some divers with established protocols. Others believe it would help yet can't encapsulate this in the form of definite procedures. Now let's consult people involved in research.

Gene Hobbs, the 2010 DAN/Rolex Diver of the Year, works at Duke University's Human Simulation and Patient Safety Center and its Center for Hyperbaric Medicine and Environmental Physiology.

"There has never been any research on this, but I did run some numbers one time. I do know a few folks who do surface O_2 for that same reason, and they report that they are happy with the procedure. That theory seemed to hold with our current risk models.

"The modern probabilistic decompression models show a reduction in DCS risk from any oxygen breathing prior to an altitude exposure for the profiles I ran. Of course, they also show a reduction from extending the in-water oxygen breathing period as well. We published the abstract showing that O_2 breathing with a last stop at 6 meters (20 feet) versus 3 meters (10 feet) does not make much difference on DCS risk, but it's hard to guess if the difference between an in-water extension versus a surface exposure would be significant. I did not run those numbers.

"But if there is enough interest, I could probably talk the Navy into letting me run more numbers through their models. I'm still not sure it's worth putting the idea in the minds of the general diving public though."

But Gene, why do you think the idea is not worth it? It saves time, which means more dives and accordingly more enjoyment and more profit for the dive operators.

"I am going to ignore the legal ramifications when you start talking about 'recommendations' that have no basis in science or in any way could be argued as less conservative. The first person to get hurt or the first flight that is diverted to facilitate treatment would be ugly.

"What it really comes down to, in my mind, is acceptable risk and should that level of risk be set by the industry or the individual. Unless there is a reason to change the current level of acceptable risk or even just redefine the current risks, nothing will change. Given the severe lack of funding for decompression research, I doubt anything would come of it unless the Special Warfare (SPECWAR) operators asked for it. That's how the current U.S. Navy flying-after-diving table came to be.

"Fun note: This data set included flights to 25,000 feet (7,620 meters) after dives. The interesting thing here is that a 25,000-foot (7,620-meter) exposure requires a 30-minute oxygen prebreathe anyway."[7]

More interesting for me, the same study concluded that the U.S. Special Operations Forces' prescribed minimum preflight surface interval of 24 hours may be unnecessarily conservative.

"The current recommendations are very conservative based partially on the concerns of a rapid cabin depressurization. I understand the concern but feel it would be useful to take another look at that given the data available. High altitude exposure has its own risk without a dive prior, but that probability of DCS given the estimated altitude and short exposure just does not seem that high, based on one evidence-based review by NASA.[8] *The U.S. Navy has reported 11 cases of DCS in one case review of 205 cases of 'loss of cabin pressure' covering the years 1969 to 1990.*[9] *The Canadian military reviewed 'loss of cabin pressure' in transport aircraft and reported no cases of DCS in 47 incidents.*[10] *Any reduction in time to fly by an O_2 breathing period would increase this risk, but is that acceptable?*

"A good example of a fun idea that was published but never turned into something more was an article in Advanced Diver *a while back on 'Benthic Mix Switching' by Glenn H. Taylor. That could theoretically do the exact same thing as your idea, since none of the tissue gas loads would be as high as they would on a 'traditional' schedule. Nothing has come of it, and I seriously doubt it ever would. Again we are back to the lack of funding.*

"I can't really say any of the available algorithms have a real advantage over the others. It could be interesting to try this with all of the algorithms, but I can't see how I would use or trust it. Until the model could also calculate a flight ascent with a max altitude of 8,000 feet (2,438 meters), any stop time would be a guess at best. Altitude decompression is a very different animal from diving decompression. That's why the Navy and Air Force have different models for each condition.

"Another consideration is that most of the divers I know who are bothered by these guidelines are active divers that are rarely out of the water for more than a couple of days. Their level of acclimatization to decompression could be protective when it comes to the additional decompression stress associated with a flight after a short surface interval. Unfortunately, there is not a good method for determining how long it would take for this acclimatization to occur or how long it takes for it to diminish. Divers never talk about 'work-up' dives anymore, but the physiology is the same now as it has been since Haldane.

"One other thing that has to be considered is repetitive flights after diving. These will also include bubble growth and reduction and changes in the gas dynamics that have never been looked at, and the current guidelines fail when it comes to this topic."

Neal Pollock is a research director at DAN and a research associate at Duke University's Center for Hyperbaric Medicine and Environmental Physiology.

"Trials were conducted at Duke University in the late 1980s to evaluate the efficacy of surface interval oxygen breathing (SIO$_2$) to reduce decompression stress. We conducted

Phase II controlled open-water trials in a Florida spring in 1991. We did have good success in extending no-decompression time with 30 minutes of oxygen breathing, but the demands of the effort were onerous (maximum of five minutes between surfacing and initiation, aggressive mask management to ensure maximal oxygen fraction breathing and absolute maintenance of rest throughout the breathing period). NOAA had funded the work with an interest in developing time-saving operational protocols, but the restrictions on postdive practice contributed to a subsequent decision to not pursue implementation.

"The effort was proof of concept. SIO₂ dive tables were not developed through the effort."

First, I tried to calculate the no-fly time after diving. Using VPM, the results implied shallower depths do not necessarily mean shorter no-fly times, yet shorter dives seem to always contribute to shorter no-fly times. Using ZH-L16, the results implied that the no-fly time is directly proportional to the depth. It also implied it is directly proportional to the bottom time.

Both algorithms resulted in less no-fly time when trimix diving was considered rather than air diving. This was expected, as the inspired helium content on the surface was practically zero, so the compartments of the trimix diver's helium would tend to offgas faster than those of the air diver. The fact that helium is a faster gas than nitrogen will only decrease its elimination time at the surface.

Accelerating the no-fly time I calculated was the next step. The time savings in the no-fly time for both air and trimix once surface oxygen breathing was engaged were almost the same. Adding several minutes to the last stop of the last dive seemed to help accelerate the no-fly time.

All in all I was not happy with this approach. Since calculating the no-fly time was not my objective, I decided to deal with it as an input. The program asked for the no-fly time the diver would normally subscribe to without taking into consideration any surface oxygen breathing. Using ZH-L16, the program calculated a corresponding gradient factor to the proposed no-fly time. For example, if I would have done these dives, I would have considered a preflight surface interval of at least 24 hours. Now we entered the 24 hours to the program as input and asked: What is the corresponding gradient factor that makes me safe to fly *not before* this preflight surface interval?

The program's accuracy of calculating the gradient factor is 1%, which means that the no-fly time I assumed as input might increase (but not decrease) a few minutes.

But what will I do with a gradient factor? I'll run the ZH-L16 after breathing the surface oxygen and utilize this gradient factor to calculate the no-fly time after breathing the surface oxygen. The time saving is the difference between the proposed input and this last output.

In bullet points, the model is as follows:

1. After planning all your dives (the model doesn't matter), the program asks for the no-fly time you would normally subscribe to. You input this value. Let's call it T1.

2. The program then runs ZH-L16 and asks it to do a deco stop at sea level to be able to ascend to the cabin altitude in a time period of T1. ZH-L16 tries with various gradient factors until it succeeds. Let's call the resulting gradient factor GF.

3. The diver breathes oxygen on the surface.

4. The program then runs ZH-L16 *with GF* to calculate the new no-fly time according to the new tissue tensions after breathing oxygen. The result is the new no-fly time, T2.

5. The no-fly time *acceleration* is T1 – T2.

The results of this new approach suggested that the time savings after trimix dives are higher than those after air dives. Does this make sense to you?

Now retired, decompression physiology expert **Michael Powell** worked at NASA as a research biophysicist in the bioastronautics division. He developed a scale for grading Doppler-detected bubbles.

> *"Data is the best way to calculate anything related to diving and decompression. Sufficient data, as you know, are not always available.*

> *"Algorithms are calculation methods and are not true physiological models. While 'model' is a term used in barophysiology, there do not exist any actual, valid models that agree with physiology in most aspects.*

> *"Bubble size is controlled, but bubble number can change (apparently) with exercise (any physical activity, such as hauling tanks, dive gear and suitcases). Nuclei concentration is not included in any model. Although number is stated as being controlled in the VPM model, I do not believe that variable nuclei number is in the VPM algorithm — except where nuclei replacement is calculated in Dr. Wienke's model.*

> *"Oxygen breathing at the surface should definitely increase safety. Your model shows some improvement, but the decreased fly time is not improved as one might guess. We used long oxygen breathing following dives in the Neutral Buoyancy Lab (testing EVA [extravehicular activity] procedures). This was to allow the astronauts to fly back to their home base with a shorter surface interval. It was found to be very successful if the surface interval was originally at least 10 hours. As long as the dive-fly interval is at least 12 hours, the diver will probably be OK. Shorter surface intervals are very sensitive to the diver's activity (hauling gear, suitcases, etc.) that can generate microbubbles in an unpredictable manner."*

Almost one year later, I decided to walk an extra mile. I have been looking at the no-fly time as a mandatory deco stop. Now why not just consider it a surface interval — a surface interval we want to accelerate, regardless of its duration? This way Ultimate Planner does not need to be fed the no-fly time or attempt to calculate it.

After any dive, we surface with some residual inert gas in our tissues. To plan a second dive, any planning tool will ask for the surface interval. Software programs capable of planning repetitive dives must predict the inert gas kinetics for each tissue compartment during this surface interval. Both VPM-B and ZH-L use the Haldane equation to do that.

The new approach I propose is to calculate the different tissue compartment gas loadings directly after the last dive, recalculate them after a surface interval on a rich mix (rather than air) and finally loop back to see how much time would be needed to reach the same loading levels if the surface interval would have been done on air. This way I don't need to calculate or assume a no-fly time. I don't need to assume a surface "dive" either. More important, the method of calculating the surface interval effect on tissue compartment gas loadings (Haldane equation) is not one of the controversial issues. If you plan repetitive dives, your planning tool — be it a Bühlmann table, a software program or a dive computer incorporating a perfusion-limited model — is already using this equation.

The results of this newer approach suggested that the time savings after trimix dives are slightly higher than those after air dives. Again, does this make sense?

Two years later, I finally realized why trimix dives contributed to better no-fly time savings. Whether the diver is breathing air, rich nitrox mix or pure oxygen on the surface, the helium gradient is at its highest, because the inspired helium content is zero. Ultimate Planner does not allow accelerating the no-fly time on any mix containing helium. Since the nitrogen gets its gradient increased, the time savings after air dives should always be higher than those after trimix dives.

It was found that, depending on the duration and timing of breathing the surface oxygen, some tissue compartments offgas nitrogen to the extent that the nitrogen pressure in these tissues became less than the inspired/alveolar nitrogen pressure of the air at sea level. This means that as soon as the surface oxygen was disengaged, these tissue compartments began ongassing nitrogen — on the surface, at sea level. Since the algorithm Ultimate Planner uses to accelerate the no-fly time compares total inert gas tissue pressures (ppN_2 + ppHe), model calibration was needed to alleviate the effect of this "counterdiffusion" on the results. This calibration was done in October 2013.

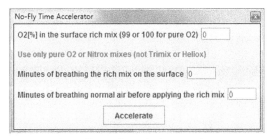

Ultimate Planner's no-fly time accelerator

The gain increases with the delay of oxygen application. Although this might seem at odds with the current understanding of offgassing gradients, it's a normal mathematical result for employing exponential equations such as Haldane's. In layman's terms, the gradient is high enough immediately postdive, so "giving it a hand" at this point by breathing surface oxygen will not have the same effect as giving it the very same hand when it is substantially lower. Not convinced? Let's take a look at some numbers.

According to ZH-L16, a diver is safe to surface with the 5-minute halftime compartment holding a nitrogen tissue tension of 29.6 meters (97 feet), which is equivalent to 2.96 bar. Let's say the diver surfaced with this compartment holding a nitrogen load of 2.76 bar (just for demonstration — that is much too high for real dives). If the diver breathes pure oxygen for 5 minutes directly after surfacing, the tissue tension will be (2.76 – 0.0 / 2) + 0.0 = 1.38 bar. After 5 more minutes but on air, the tissue tension will be (1.38 – 0.76

(alveolar ppN$_2$) / 2) + 0.76 = 1.07 bar. On the other hand, if the diver delayed the surface oxygen for 5 minutes, the tissue tension would be (2.76 – 0.76 / 2) + 0.76 = 1.76 bar just before the oxygen application and (1.76 – 0.0 / 2) + 0.0 = 0.88 bar after breathing it for 5 minutes. So delaying the surface oxygen actually boosted the offgassing.

Note: Almost one year after I published my inaugural editorial about accelerating the no-fly time in the first issue of *Tech Diving Mag* (December 2010), DAN started an FAD calibration study. The study took place at the Duke hyperbaric center, and one of its objectives was to investigate the possibility of decreasing the preflight surface interval by breathing oxygen at the surface. The results and conclusions of the study have not yet been published.

Diving at altitude

As you go above sea level, the ambient pressure decreases. To adapt to the new ambient pressure, your body requires some time (at least 12 hours after reaching the desired altitude, 24 hours is common practice). This adaptation time is to release the excess nitrogen from your body, as your tissues' nitrogen loading are now *not at equilibrium* with (actually higher than) the new, reduced ambient pressure at altitude. If you dive without considering this adaptation time, the situation would be similar to making a repetitive dive after decompression from a saturation dive at 1 ATA. Even after the diver is acclimatized, the decompression time for a given dive at altitude is still longer than that of a corresponding dive at sea level because bubbles grow larger at reduced barometric pressure. Accordingly, expect the NDLs of altitude dives to be shorter than those of the corresponding dives at sea level.

For planning dives at altitude, you can use a special set of tables, computer software, or dive computers. Your planning tool should account for how far you are above sea level and how much time you have spent at this altitude. So if a dive computer is your choice, the instrument must travel with you at all times, and the instructions for use at altitude must be followed. If you are using a software planning tool, make sure to adjust the altitude settings.

If you are using a set of tables, check the ascent speed the tables recommend. Some tables recommend slower ascent speeds for diving above sea level.

One final note: Diving at altitude and FAD are two very different issues. I frequently get asked about planning a sea-level dive as if it is at cabin altitude, doing the required decompression accordingly (obviously more decompression time than at sea level), then flying directly without a preflight surface interval. The answer is no, don't do that.

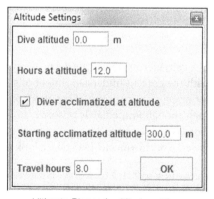

Ultimate Planner's altitude settings

Conservatism

Individuals vary in their response to the inert gas dose, and the difference can be quite large. Powell elaborates: *"It is of interest to report here a Russian study partly inspired by the ARGO tests. A group of 34 individuals were depressurized in a chamber to high altitudes (not all at the same time) until some got joint-pain decompression sickness. These were removed, and a few days later the group was decompressed to a higher altitude. As subjects got DCS, they were removed from the group until all had finally succumbed to joint pain. The distribution of subjects formed a normal S-shaped curve. The difference between the 'sensitives' and 'resistants' was quite large in terms of susceptible pressure. This illustrates what has been strongly suspected for decades."* [11]

Accomplished divers often advise not to push a model to the edge of its limits. Use some degree of conservatism.

To add a safety factor to raw dissolved-gas decompression models, software developers used to increase the inert gas loadings, the inert gas content, the bottom time or the depth. For example, depending on the software tool, a 10% safety factor would multiply all the tissue loadings by 10% at the end of the bottom time, increase the inert gas content in the breathing mix by 10%, increase the bottom time by 10% or increase the depth by 10%.

Most of raw dissolved-gas decompression models result in very shallow stops. The safety-factor approach lengthens the stops but does not introduce deeper ones. Based on anecdotal data, divers agreed it would be better to start deco deeper, so they used to insert arbitrary deep stops. The two main approaches for doing that were Pyle's and GVE. Some software developers invented yet another approach: add a very fast tissue compartment. Usually the halftime of this compartment is in the 1- to 2.5-minute range (for nitrogen). This ultrafast compartment would control the ascent and introduce deeper stops.

Baker's gradient factors virtually killed all the former approaches. The low factor introduces deep stops, and the high one lengthens the shallow ones.

As discussed in Chapter 6, to control bubble growth VPM-B assumes initial critical radii, one for nitrogen and the other for helium. Only bubbles with *bigger* radii than these critical values are allowed to grow. The current default initial critical radii values that we feed the VPM-B algorithm are 0.55 and 0.45 micron for nitrogen and helium respectively.

To generate more conservative schedules, VPM-B assumes bigger initial critical values. For example, 22% conservatism means 0.55 * 1.22 = 0.671 micron for nitrogen and 0.45 * 1.22 = 0.549 micron for helium. Now VPM-B will need to control more bubble sizes to not allow them to grow, which in turn will result in more conservative schedules.

Now divers usually use more than one model to plan their dives. They frequently double-check their plans and compare the generated schedules using different algorithms. They often compare the VPM-B-generated schedules with those of Bühlmann's ZH-L16 with gradient factors. This usually works fine in a range of depths and bottom times. With the dives getting bigger every day, the VPM-B plans cannot cope with the extended shallow stops the ZH-L16 advises.

To add a second dimension of conservatism to VPM-B, software developers started implementing model variations, the most popular of which are Hemingway's /E (2005), Shearwater's /GFS (2011) and my /U (2011), which were discussed in Chapter 6.

Ultralong halftimes

Since all tissues are alive, they need oxygen and nutrients, and they produce CO_2 as a waste product. Haldane assumed three hours are enough for nitrogen loading to reach saturation in goats and presumed humans would reach full saturation in five hours. This means that the slowest compartment halftime would be around 50 minutes.

The perfusion-limited decompression models use sets of halftimes covering a much wider spectrum. For example, the halftime of ZH-L16's slowest compartment is more than 10.5 hours. It is established that a tissue with such a limited blood flow to create such a long a halftime would not have a reasonable blood supply (i.e., oxygen) to be viable, except for teeth and the very solid portions of bones, which are not really capable of dissolving inert gas.

If these ultralong halftimes are not physiological, what would they be representing? One opinion is that they are actual gas bubbles in equilibrium with the tissues and that the compartments represent the different portions of the tissues and even portions of the cells, including the would-be existing bubble seeds. In this context, dissolved-gas models are actually handling bubbles by accounting for their influence on inert gas uptake and elimination rates rather than trying to control their size and number.

Now let's look at what ultralong halftimes would do to VPM-B-generated schedules. The majority of VPM-B planning tools use the ZH-L16 compartment set advised by Bühlmann. For the technical dives we do today, even the bigger ones, using a different set covering a wider range of halftimes will not affect the generated schedules in any tangible means. The number of compartments will not affect the generated schedules either, as long as they are distributed properly. But there would be a significant difference with the extremist of dives.

Dec-12 is an alternative compartment set implemented in Ultimate Planner. It consists of 12 pairs of compartments distributed evenly over a wider range than that of the ZH-L16. The diver can choose which set to employ. The generated schedules are generally very close, but take for instance Ben Reymenants' 239-meter (784-foot) dive in Sra Keow cave near Krabi, Thailand. On February 18, 2007, Reymenants glimpsed Sra Keow's floor after a 25-minute descent. For this dive, using Dec-12 instead of the ZH-L16 halftime set would result in some 30-minute increase in the total run time. If, on the other hand, we used VPM-B with a set of halftimes covering up to 200 minutes only (approximately the slowest *physiological* tissue halftime, fatty marrow), Reymenants' schedule would have been cut by some 200 minutes.

Asymmetric gas kinetics

In 1960 Hempleman presented evidence from animal studies that the uptake and elimination of inert gases were not symmetrical.[12] He illustrated that body tissues absorbed inert gas more quickly than they eliminated it. In 1969, to fix the rate of elimination, he suggested a default value of 0.67 multiplied by the rate of absorption.[13] Based on observations from physiological experiments, Berghage, Dyson and McCracken concluded in 1978 that the rate of uptake and elimination of inert gas was not the same.[14] Experiments of Rogers and Powell in 1988 confirmed this concept.[15]

Several decompression models employed this principle and were implemented in dive computers. This was done in one of three ways. The first was to assume higher halftimes for the elimination process (ongassing halftimes are lower that offgassing halftimes). The second was to implement an exponential-linear model rather than the original Haldanean exponential-exponential model. The third was having a symmetry variable multiplied by the rate of absorption.

The /U model variation implemented in Ultimate Planner uses the third method. Setting this variable to 100% takes you back to the original VPM-B results. The different degrees of asymmetry are 100% (symmetrical), 95%, 88%, 78% and 67% (the value Hempleman suggested in 1969).

Oxygen bends

This topic is frequently raised on Internet forums. Can one get bent because of oxygen instead of inert gas? Powell provides the answer:

> "Something I have been asked about over the years is a curious phenomenon called 'oxygen bends.' This was first seen by K.W. Donald in goats in a study conducted in 1955. Donald took goats to depth in a hyperbaric chamber using air as the compression gas. He found the specific depth for the appearance of limb bends and then reduced the nitrogen pressure by a few psi (1 bar = 14.5 psi). On another day, he then added several psi of oxygen to the air and dived the animals again. He noticed that the goats were badly afflicted with DCS, staggering badly and falling down, and almost to the point of death with much labored breathing. Within about five minutes, however, the goats stood up, walked and recovered with no residual signs or symptoms.

> "This is what I believe occurred. With a very good dose of dissolved nitrogen [and oxygen], there was a considerable bubble load as the tissues dumped numerous bubbles into the vena cava and the heart. This formed a right heart airlock, and there was little cardiac output. That is, the heart was unable to pump blood, since it contained so many bubbles. The heart was simply compressing foam. Little blood flowed to the brain.

> "In a few minutes the body metabolized the oxygen, and the bubble load was considerably reduced. With the reduction of foam in the right heart, the cardiac output was increased and blood flow again to the lungs and brain. Such could never happen if the breathing mix was primarily air, since the nitrogen in the tissues, and later bubbles, could never be metabolized.

> "While this is a very specialized instance of a rapid decompression at a DCS limit, nevertheless, it does show that such a case of DCS is possible. This is not a situation

typically encountered in normal scuba diving operations but does illustrate a point in a laboratory situation."[16]

In the mid-1980s, animal experiments suggested high O_2 can increase the risk of DCS. Verification for its effect on human decompression was needed. In 1985 a study with a total of 477 controlled dives was performed to assess the relative risks of O_2 and N_2 in human no-stop dives.[17] Statistical analysis by maximum likelihood estimated that O_2 had zero influence on DCS risk, although data variability still allowed a slight chance that O_2 could be 40% as effective as N_2 in producing a risk of DCS. The final conclusion was that consideration of only inert gases was justified in calculating human decompression tables.

Patent foramen ovale (PFO)

Remember the "undeserved hit" we referred to in the introduction of this book? Here is one possible reason behind such an unfortunate event.

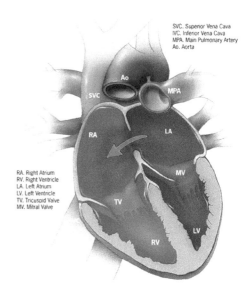

A human heart with left to right shunting
Photo courtesy of Wikipedia. Public domain. "Atrial septal defect."

As the lungs are unusable before birth, the oxygenated blood comes directly from the mother, bypasses the lungs and passes from the right atrium (venous side) to the left atrium (arterial side) through a hole called the foramen ovale.

At birth, pulmonary blood flow significantly increases. Blood rushes through the lungs, increasing the left atrial pressure. The resulting atrial pressure difference compresses the foramen ovale and *functionally* closes it. Complete sealing of the foramen ovale usually occurs later in infancy in most people. In some 25% of the general population, however, the state of complete sealing is not achieved. This type of *atrial septal defect* is known as PFO, and it leaves the sufferers at risk of *shunting*. Shunting is when the blood flows directly from the left side to the right side of the heart, or vice versa. Under normal conditions, this does not constitute a health hazard. For divers, the left to right shunting in itself does not constitute DCS-related problems. On the other hand, if right to left shunting occurs, venous blood containing inert gas bubbles may bypass the lungs (where it should be filtered) and pass to the arterial side to be pumped around the body.

Powell provides this account on PFO-related studies conducted at NASA:

"In decades past, bubbles were thought to move from the venous to the arterial side primarily through the blood vessels of the lungs ('arterialization'). I believe (from my

early studies in the 1970s) the majority of gas bubbles in the venous return have their genesis in muscle and adipose capillaries and then are released, especially during muscle contraction, into the central venous return. Studies by other researchers showed that vasodilators (blood vessel expanders) would allow bubble 'spillover' at the lungs in anesthetized dogs.

"Other researchers had measured mean right side and left side atrial pressures (in the heart) in anesthetized pigs. Pressure measurements did not give an indication that right (venous side) was increased over left (arterial side) when arterialization occurred with the relatively large gas loads employed. However, at some point in the cardiac cycle, the gradient did reverse, if only momentarily, with these large gas loads. Unfortunately, these reversals can be caused by the Valsalva maneuvers or something similar. We did not get the chance to study this in the detail while I was at NASA, but I did study the medical literature and noted reports of 'Valsalva-like maneuvers' following activities such as straining, coughing, sneezing, pulling, tugging and the like. This is very important for scuba divers since a slight breath hold while climbing a boat ladder is a Valsalva maneuver in everything but name. Yes, a rose by any other name is still a rose, as Shakespeare once said.

"As a part of our ARGO decompression studies at NASA, transcranial Doppler (TCD) ultrasonography of the cerebral arteries was performed in the hypobaric chamber. TCD is noninvasive and gives unequivocal evidence of the presence of gas bubbles in the arterial system of the brain. Monitoring was performed with TCD on individuals with microbubbles on precordial Doppler at the conclusion of a three-hour exposure to altitude during the ARGO series. We chose the end of the altitude run to eliminate the need to repressurize the subject when brain bubbles occurred (they were repressurized anyway). We wished to know if DCS occurred with brain bubbles present it would have appeared and repress was just moments away at the end of the experiment. (This satisfied the ever-vigilant Human Subjects Committee.) Comparisons made between: (i) ground-level saline-contrast echocardiography for PFO, (ii) precordial Doppler severity (Spencer grades) and (iii) cerebral artery Doppler signals during hypobaric exposure.

"Of the individuals who generated decompression gas bubbles (18 in our studies), two had a resting PFO and only one gave evidence of arterialization at the end of the three-hour hypobaric exposure. These two individuals displayed 'resting arterialization,' that is, a Valsalva augmentation was not needed to demonstrate the patency in the ground-level test. One, surprisingly, did not evince arterializations in the hypobaric chamber even with Spencer grade IV bubbles (the highest grade). From this limited database, a patency without provocation and a high Spencer grade appears to be a necessary but not sufficient condition for arterialization.

"We were not able to secure funding for further studies, but divers should be aware that a PFO is not 'the kiss of death.' Many divers want to get a PFO test, not realizing that a PFO appears to be of little concern for DCS, and the test is expensive and not entirely without risk. In addition, bubbles can arterialize not only through a PFO but also through the pulmonary capillaries." [18]

In 2003 a study was conducted to assess the risk of decompression illness (DCI) in relation to the presence and size of PFO.[19] DCI covered the combination of DCS and arterial gas embolism (AGE, caused primarily by lung overexpansion). The study, with a sample size of 230 divers, concluded that the risk of suffering DCI in divers with PFO is five times as high as in divers without PFO. Another conclusion was that the risk of suffering a major DCI parallels PFO size.

Multilevel dives

Imagine a cave with several "jumps" or a shipwreck in a straight position, and you want to take a look at its propeller and upper deck in the same dive. Hence emerges the need to abandon the square profile and plan multilevel dives.

Multilevel dives are three types: multilevel descents, multilevel ascents, and recompression (sawtooth) profiles. Multilevel descents are usually accompanied with gas switches to ensure the gas in use is breathable at every phase down until the desired depth. An example is to use trimix (20, 25) to 60 meters (200 feet), switch to trimix (12, 53), then descend to 100 meters (330 feet). Decompression models should have no problem dealing with multilevel descent profiles.

Multilevel ascents are usually more complicated. When you insert a level shallower than the previous one but still deeper than the start of the decompression zone, there's no problem. On the other hand, when you insert a level shallow enough to be in the decompression zone, you're actually adding extended decompression stops the model didn't advise. However, deco-planning software tools can deal with your intervention by incorporating a simple multilevel algorithm that will adjust one of the stops to cope with your input. An example is to dive to 60 meters (200 feet) for 30 minutes, and on your way up you stay for 20 minutes at 25 meters (82 feet). The software will simply shift the 24-meter (80-foot) stop down a little bit to meet your new desired level (25 meters or 82 feet), extend the stop to meet your new desired segment time (20 minutes), and go on with the rest of the schedule (taking into consideration the gas uptake and elimination that took place at that stop).

Ultimate Planner 1.5 by Asser Salama.

Warning: This software is intended for demonstration purposes only. The author accepts absolutely no responsibility for the schedules generated by this software. Use it at your own risk.

VPM-B/U: OFF
Tissue compartment set: Dec-12
Conservatism: 0%
Altitude: 0.0m
Leading compartment enters the decompression zone at 44.5m
Run time includes the ascent time required to reach the stop depth

Depth	Seg. Time	Run Time	Mix	ppO2
60.0m	27.0	(30)	Tx20/25	0.20 - 1.40
33.0m	1.0	(34)	Tx20/25	1.40 - 0.86
30.0m	1.0	(35)	Tx20/25	0.86 - 0.80
27.0m	1.0	(36)	Tx20/25	0.80 - 0.74
25.0m	20.0	(56)	Nx40	0.74 - 1.40
21.0m	1.0	(57)	Nx40	1.40 - 1.24
18.0m	1.0	(58)	Nx40	1.24 - 1.12
15.0m	1.0	(59)	Nx40	1.12 - 1.00
12.0m	1.0	(60)	Nx40	1.00 - 0.88
9.0m	4.0	(64)	Nx40	0.88 - 0.76
6.0m	20.0	(84)	Oxygen	0.76 - 1.60

OTU of this dive: 130
CNS total: 81.0%

3462.8 ltr Tx20/25 ------> (5194.2 ltr for thirds)
1324.4 ltr Nx40 ------> (1986.6 ltr for thirds)
495.0 ltr Oxygen ------> (742.5 ltr for thirds)

A multilevel ascent dive schedule

The big issue exhibits itself when you add a shallow level that lies in the decompression zone and then add a deeper level. This type of recompression is not recommended by the existing decompression models. Actually some of them are not designed to deal with it, because the way they treat the compression segments is totally different from how they decompress. Once the diver enters the decompression zone, the way the model behaves takes a different turn. Going back would be unexpected.

Decompression-planning software tools tend to overcome this problem by computing an intermediate profile as if the dive would end in the water, without taking the recompression segment into consideration. Then a new descent for the recompression part is added as if it is a new dive starting from within the water column. It is simple programming — just another nested loop. The problem with that approach as it relates to decompression is that when it comes to dual-phase models, some variables need to be initialized at the beginning of every dive. Considering the dive as two separate dives with two different descents from different depths would upset the model. So far, decompression-software programmers have not agreed on which variables to recompute and which ones to reset or even turn off. For example, some VPM-B advocates argue that the critical volume algorithm needs to be turned off, otherwise the ascent plan will not be safe enough. Others believe that the most significant part is to recalculate the maximum crushing pressures on the second descent while resetting the allowable gradients to their initial values.

Planning recompression dives is less of an issue with dissolved-gas models, which are much simpler than dual-phase models, and there are fewer tricky parts to deal with.

I think that dual-phase models should not be "taken by surprise" — they need to know all details beforehand. For example, let's consider a 30-minute dive to 45 meters (150 feet) on air. Using VPM-B with no conservatism, the total run time is 105 minutes. To accelerate the decompression, we'll use EAN80 starting at 9 meters (30 feet). The total run time becomes 68 minutes. Now take a closer look at the two plans.

You might wonder why these VPM-B schedules are not exactly the same up until the 9-meter (30-foot) mark. Isn't that the point where the gas switch took place? In both plans we are using air all the way down and then up to 9 meters (30 feet), so why does the difference in run time start at 21 meters (70 feet)?

Ultimate Planner 1.5 by Asser Salama

Warning: This software is intended for demonstration purposes only. The author accepts absolutely no responsibility for the schedules generated by this software. Use it at your own risk.

VPM-B/U: OFF
Tissue compartment set: ZH-L16
Conservatism: 0%
Altitude: 0.0m
Leading compartment enters the decompression zone at 33.0m
Run time includes the ascent time required to reach the stop depth

Depth	Seg. Time	Run Time	Mix	ppO2	Depth	Seg. Time	Run Time	Mix	ppO2
45.0m	27.8	(30)	Air	0.21 - 1.16	45.0m	27.8	(30)	Air	0.21 - 1.16
24.0m	1.0	(33)	Air	1.16 - 0.71	24.0m	1.0	(33)	Air	1.16 - 0.71
21.0m	2.0	(35)	Air	0.71 - 0.65	21.0m	1.0	(34)	Air	0.71 - 0.65
18.0m	3.0	(38)	Air	0.03 - 0.59	18.0m	3.0	(37)	Air	0.03 - 0.59
15.0m	5.0	(43)	Air	0.59 - 0.52	15.0m	4.0	(41)	Air	0.59 - 0.52
12.0m	5.0	(48)	Air	0.52 - 0.46	12.0m	6.0	(47)	Air	0.52 - 0.46
9.0m	9.0	(57)	Air	0.46 - 0.40	9.0m	4.0	(51)	Nx80	0.46 - 1.52
6.0m	48.0	(105)	Air	0.40 - 0.34	6.0m	17.0	(68)	Nx80	1.52 - 1.28

OTU of this dive: 41
CNS total: 15.2%

4596.3 ltr Air ------> (6894.45 ltr for thirds)

OTU of this dive: 72
CNS total: 28.0%

3122.6 ltr Air ------> (4683.9 ltr for thirds)
536.3 ltr Nx80 ------> (804.45 ltr for thirds)

Dive schedules demonstrating a VPM-B limitation

Using air only will result in a longer schedule, and this in turn may lead to lower allowed supersaturation gradients. Subsequently, these lower supersaturation gradients may cause the decompression to start deeper, the deep stops to be longer or both. This is one of the drawbacks of employing the critical volume algorithm in VPM-B, but not using it would result in some 20-minute increase in the total run time. It is a trade-off that VPM-B adopters have to live with, even when they are planning simple things such as lost deco gas scenarios. Planning recompression dives would be much trickier.

Of course, the problem still exists and actually would increase with real-time programs (such as the ones implemented into your dive computer). Recompression dives might cause your dive computer to behave unexpectedly, especially if it is running a dual-phase model.

Temperature

Anything that increases ongassing or decreases offgassing can contribute toward getting hit. As the body cools down, the peripheral blood vessels shut down, thereby reducing the inert gas uptake and release rates. This means that gas-exchange kinetics in tissues involved in DCS are slowed by vasoconstriction during cold exposure. Similarly, the same gas-exchange kinetics are accelerated by vasodilation during warm exposure.[20] Thus, the diver's thermal status during different phases of the dive can greatly influence the susceptibility to DCS. This means that cold conditions during the bottom time and warm conditions during decompression would be optimal for minimizing DCS risk.

In 2007 a study at NEDU concluded that divers should be kept cool during the bottom time and warm during subsequent decompression.[21] The same study found the beneficial effects of warm conditions during decompression were more pronounced than the deleterious effects of warm conditions during the bottom time. A previous study concluded divers who are warm at depth face an increased risk of DCS because vasodilation in warm divers may result in more rapid ongassing of some tissues. It suggested that a full evaluation of DCS risk should consider physiological and physical effects of ambient temperature.

The bottom-time phase of the dive usually involves some work, so it is comparably warmer than the decompression phase, where the diver is often at rest. The warmer conditions at depth may accelerate gas uptake, whereas the colder conditions during decompression would reduce gas elimination. Of course this is not good for decompression, and that is why some models compensate for this situation by generating a more conservative profile for colder dives. These models are frequently referred to as adaptive models. Ultimate Planner's /U model variation deals with colder dives. Also the ZH-L16D model variation that was implemented in 2013 is addressing the same situation.

In 2008 an interesting study was conducted to investigate the influence on bubble formation of exposure to heat *before* diving.[22] One hour after concluding a 30-minute far infrared-ray dry sauna-induced heat session at 65°C (149°F), 16 divers were compressed in a hyperbaric chamber to 30 meters (100 feet) for 25 minutes. The same dive was performed five days earlier without the sauna session. Precordial Doppler monitoring detected circulating venous bubbles. Body weight measurements were taken before and after the sauna session. The results suggested that the sauna session led to an extracellular dehydration, resulting in 0.6% body weight loss along with a significant reduction in bubble formation. The study concluded that a single predive sauna session significantly decreases circulating bubbles (probably due to sweat dehydration, among other factors) and that this course of action may reduce the risk of DCS.

The solubility of inert gas is inversely proportional to the temperature. This means the tissues will hold less gas in solution as they get warmer, which in turn could promote bubble formation or growth. This fact may explain the anecdotal observations that hot showers after diving or during surface intervals precipitate DCS.

Dehydration

Associating DCS with dehydration is as old as World War II, when it was reported as a factor that increases the risk of DCS in aviators. The mechanism was not clear at that time, and unfortunately it still is not.

Our current understanding is that when a considerable portion of the fluids leaves the body, the body concentrates the rest of the fluids internally, and peripheral flow is cut down. This would reduce the offgassing capability of the tissues, as the capillary gas exchange at the extremities becomes less efficient due to the decreased circulation. So far, this hypothesis is not established.

A study was conducted in 2007 to investigate whether hydration 90 minutes before a dive could decrease bubble formation.[23] It concluded that predive oral hydration using a saline–glucose beverage decreases circulatory bubbles, which in turn may reduce the risk of DCS.

Exercise

A study in 2009 concluded that signals consistent with microbubbles have been detected in the legs of normal human subjects after exercise.[24] The results were not new; this has been known for two decades.

In 1989 Powell submitted a proposal to NASA to run decompression-related studies.[25]

"After considerable preparations, the first depressurization commenced in July 1991. The idea is that tissue micronuclei have a limited lifetime, and they are continuously regenerated. If you do not walk, the bubbles will not be formed. As hours pass, the nuclei will shrink.

"The project plan consisted of twenty subjects in a crossover design. The subjects were [condition A] either to walk from exercise station to station or [condition B] be lying in a bed with the exercise stands nearby. Some subjects were adynamic in the first series and then later were switched into the ambulatory group. The other half were ambulatory then adynamic.

"The Doppler bubble results were definitive. In both groups, bubbles from the arms were about equal; however, leg bubbles in the adynamic group were considerably reduced. This indicated that something to do with adynamia played an important role. There was no way to tell that the effect was actually caused by a reduction in nuclei since there is not easy way to determine this. There could have been some biochemical factor, although I do not know what this might be.

"In any case, exercise was demonstrated to be a large factor in DCS bubble formation. What we determined can be applied to recreational divers, namely that physical activity (musculoskeletal stress) was a large factor in influencing the risk of decompression sickness.

"DCS is a multifactorial event, with pressure and bottom time being the primary factors. Exercise is a big one also, and aspects of this have been known since studies of 'flyers bends' during World War II. Yet we still hear of 'hydration' as a factor. Not really. Certainly it plays some role, but in most divers this has not been shown to be a big player.

"When looking at the observable effects of strenuous activity, it is clear that recreational divers should avoid heavy physical activity both in the interdive period and at the end of the dive day. I have heard of a diver who thought that he got DCS from climbing a hill that was 137 meters (450 feet) in height. He though it was an effect of altitude. No, it was from the work of the climb itself. Divers should avoid climbing onto a boat with all of their gear. This is strenuous, produces bubbles and can lead to Valsalva-like maneuvers. We have seen test subjects develop bubbles with a maneuver (pulling themselves with their arms) that immediately caused the evolution of bubbles from that arm."

Does this mean that exercise *during* decompression enhances the elimination of inert gas? A study in 2000 suggested *mild* exercise during decompression would enhance inert gas elimination and, subsequently, would reduce the risk of DCS.[26] The study concluded that it is not determined whether or not mild exercise may enhance inert gas elimination to the point of reducing the times required for inactive decompressions. Mild exercise during decompression is more of an ad-hoc practice that, as per our current understanding, will not cause any harm.

Omitted decompression

In the introduction of this book, I recalled a firsthand experience of particular importance, a case of almost half an hour of omitted decompression where the diver neither showed signs nor developed symptoms of DCI. As mentioned before, DCI covers the combination of DCS and AGE.

The diver did nothing, although oxygen was available onsite (at the surface). While the benefits have not been measured yet, under normal conditions breathing oxygen would certainly cause no harm. However, as the story goes on, we discovered it was late October yet still uncomfortably hot. Hot weather would have the same effect of a hot shower. It would contribute to vasodilation and subsequently enhance peripheral circulation, which might facilitate offgassing. Is that good or bad? It depends on the heat stress and the inert gas load. A large inert gas load could prove problematic. The solubility of inert gas is inversely proportional to the temperature. This means the tissues will hold less gas in solution as they get warmer, which in turn could promote bubble formation or growth. Exposure to heat is not what you would be looking for in an omitted decompression situation (especially given that development of DCI symptoms immediately following a hot shower has been reported and documented, even without omitted decompression).

In conclusion, not breathing oxygen under this heat stress appears sage, especially because oxygen breathing would need to be prolonged (at least 30 minutes) to be of real value. The effect of the cold shower the diver took after rinsing the gear could not be verified either. That is just an ad-hoc DCI prevention technique that, as per our current understanding, will not cause any harm in that case, especially if the shower duration is short enough not to cause an undesirable core temperature reduction.

Another unverified yet harmless DCI prevention technique the diver employed was drinking water on the way back to the dive center, because dehydration is a theoretical risk factor in DCI.

As demonstrated earlier in this chapter, strenuous activity after a dive has been shown to contribute to DCI. Our diver did not mention any such activity (other than rinsing gear, which I do not consider strenuous). The nap the diver took after the cold shower decreased the activity level to zero. Although DCI symptoms might develop during sleep, the nap the diver took was described as short. Napping has been found to be physiologically beneficial; it can help refresh the mind and improve overall alertness, which is desired in our case. One preventive DCI measure the diver did not subscribe to is to not dive for at least 24 hours. Instead, the diver crossed considerably high altitudes to go diving elsewhere. While the decisions the diver made will not contribute to DCI prevention, the diver felt that the "24-hour sit-down rule" before diving or crossing high altitudes is purely subjective (although combining both would increase the risk of DCI). The diver's approach here was clearly based on self awareness.

Should the diver have re-entered the water to complete the omitted decompression? This is a controversial issue. Some hyperbaric experts recommend that an asymptomatic diver should re-enter the water and complete the stops. Their argument is that this approach would be better than waiting for DCI to hit. Some protocols were developed for this situation, the most popular of which are from the U.S. Navy.[27] Their protocols depend on the time it takes the asymptomatic diver to return to the water and the depth of the missed stops. If the missed stops are 9 meters (30 feet) or shallower and the asymptomatic diver can return to the water within one minute maximum, then he should return to the required deco-stop depth, increase that stop time by one minute and carry on with the rest of the decompression schedule without further changes. If the missed stops are 9 meters (30 feet) or shallower but the asymptomatic diver cannot return to the water within one minute and a recompression chamber is not available onsite, he should return to the required deco-stop depth and carry on with the decompression schedule but multiply every stop time by 1.5. For omitted decompression stops deeper than 9 meters (30 feet) where a recompression chamber is not available onsite, the asymptomatic diver should return to the depth of the first decompression stop (not to the depth of the first omitted stop) and follow the original decompression schedule until the 9-meter (30-foot) mark. At 9 meters (30 feet), the diver should shift to oxygen (if available and CNS status permitting) and complete decompression from 9 meters (30 feet) by multiplying all the remaining stop times by 1.5.

Currently we do not have a standard textbook advice for omitted decompression situations when no DCI signs or symptoms manifest themselves. Some experts argue that breathing oxygen on the surface is preferred to re-entering the water and completing the omitted decompression on air.[28] The term *decompression* includes the no-stop dives as well. As the ascent rate is an essential input for generating *any* schedule (not only those dedicated to planned-stop dives), rapid ascents should always be dealt with as omitted decompression situations.

In 2009 a study was conducted to investigate the influence of in-water oxygen breathing on bubble formation following a provocative dive by comparing the effect of postdive hyperbaric versus normobaric oxygen breathing on venous circulating bubbles.[29] The result was that breathing oxygen at a 6-meter (20-foot) depth dramatically suppressed circulating bubble formation with a bubble count significantly lower than for oxygen breathing at the surface. The study concluded that this could be used in situations of interrupted or omitted decompression, where the diver returns to the water to complete the missed decompression prior to the onset of DCI symptoms.

In-water recompression (IWR)

The difference between IWR and omitted decompression protocols is that IWR is the term used for treating divers with signs or symptoms of DCI, whether they miss stops, violate ascent rates, or just get an unearned hit. You may frequently read that a diver who develops symptoms of DCI should not re-enter the water no matter what and should breathe surface oxygen instead until either all oxygen supplies are depleted or the emergency medical services personnel (which should be contacted as soon as possible) arrives, whichever comes first. The available data clearly support the use of surface oxygen for symptomatic cases.

On April 6, 1994, cave-diving pioneer Jim Bowden made an attempt to bottom out El Zacatón sinkhole in Mexico. A gas shortage forced him to turn his dive at 282 meters (925 feet). Upon surfacing, he suffered a left shoulder DCS hit and was treated onsite with IWR. Clearly, if a portable recompression chamber like the one used by Jacques-Yves Cousteau and his team in their epic expeditions on the *Calypso* had been available onsite, Bowden would not have used IWR.

In conclusion, IWR should be considered an option of last resort for early management of significant symptoms of DCI. If, and only if, the incident involves a group of high-profile technical divers, appropriately equipped for in-water recompression where no recompression chamber is available onsite and there is no prospect of reaching a hyperbaric medical facility within a reasonable timeframe, the edict against re-entering the water could be relaxed. To get a better picture of what "appropriately equipped for in-water recompression" means, imagine a dive expedition in an isolated location where the team has installed a decompression stage or an underwater habitat along with a system for delivering pure oxygen underwater. Full-face masks with communication capabilities if a habitat is not used are very strongly encouraged. Having adequate thermal protection, sufficient volumes of gases (preferably pure oxygen), a tender to escort the victim, and a valid IWR treatment schedule is compulsory.

The U.S. Navy's IWR treatment protocol states that the stricken diver should begin breathing pure oxygen immediately at the surface for 30 minutes before committing to recompress in the water. If the symptoms stabilize, improve, or relieve, the diver should not attempt IWR unless the symptoms reappear with their original intensity or worsen when oxygen is discontinued. Breathing pure oxygen should continue at the surface as long as the supplies last, up to a maximum time of 12 hours. The victim may be given air breaks as necessary. If surface oxygen proves ineffective after 30 minutes, IWR should begin. The IWR treatment schedule depends on the gases available and the DCI symptoms present. With symptoms such as unconsciousness, paralysis, vertigo, respiratory distress and shock, the risk of increased harm to the victim from IWR probably outweighs any anticipated benefit. In general, victims experiencing some of these symptoms should not be recompressed in the water. Instead, they should be kept at the surface on pure oxygen, if available, and evacuated to a hyperbaric medical facility regardless of the delay.

Another option some expedition leaders prefer to IWR is saline intravenous therapy along with pure oxygen at the surface and some pain-relief medications. A newer yet untested approach is to administer intravenous perfluorocarbons (PFC) because of their capability to dissolve vast amounts of polar gases. In 2008 a doctoral thesis concluded that the administration of intravenous PFC emulsions reduces both morbidity and mortality of DCS.[30] The results illustrated there is an 11% increase in the arterial oxygen content over the saline control. However, these are animal results. Tests on human subjects have yet to be conducted.

Acclimatization

A long time ago I was taught that multiday diving is a significant risk factor for DCS. Some dive computers available today penalize the diver by dialing down the M-values through a reduced gradient. For example, the Suunto RGBM will calculate some 90% reduction for multiday diving in a seven-hour surface interval.[31]

Acclimatization is when a diver is at *reduced* risk of DCS as a result of conducting dives during the preceding days. In 1967 a study demonstrated the effect of acclimatization in an analysis of 40,000 air decompressions of caisson workers. The incidence of DCS dropped from approximately 12% to 1% over the first 10 to 15 decompressions (5 days per week). Acclimatization was lost during two to 10 days off.[32]

Through the one-year period March 4, 1989, to March 4, 1990, Gilliam compiled data for a total of 77,680 dives, including customers and professional staff aboard his 140-meter (457-foot) diving cruise ship *Ocean Quest*. He noticed that some validity to the hypothesis of what he called then "adaptation" must be given serious consideration. His team of dive professionals worked aggressively for four straight days and then received three days off before resuming the same schedule. Most made between 500 and 725 dives in the 1-year period. Many routinely performed dives in the 75-meter (250-foot) range or greater, on air, with subsequent repetitive dives, and yet no DCS hits were recorded in any staff.[33] Gilliam suggested that the "multiday skip" protocol should be validated later.

Some interesting data on acclimatization was presented at DAN's Technical Diving Conference in January 2008. The general viewpoint endorsed was that multiday diving was not a significant risk factor for DCS.[34] In 2013 a study showed that four consecutive days of identical daily diving could *reduce bubble formation*, representing what was likely a *positive acclimatization* to diving.[35] This study pointed out that although bubbles did not equate to DCS, it was reasonable to accept the presence of fewer bubbles was desirable. So in conclusion, the mechanism of acclimatization to decompression stress is still unknown.

Washout treatment

It is not unusual for dive professionals at tourist resorts to do repetitive dives over extended periods, especially in high seasons. This is also true for busy dive instructors who prefer conducting their courses in tropical locations, for example those who fly from Europe and Asia to the Red Sea with their students. Those dive professionals frequently do several dives to various depths every day, with very little surface interval in between.

There is a misconception that such practice would result in a degree of tissue inert gas saturation. Divers who believe that also believe that the approach to deal with it is a "washout dive," which is a hyperbaric chamber dive to washout any excess tissue inert gas loadings.

I do not know the origin of this misconception. As far as I am aware, there is nothing published in reputable magazines or credible journals regarding this fallacy. The concept is at odds with our current understanding of decompression. As we have already seen, there are two factors contributing to the onset of DCS: dissolved gas and bubbles. Doppler tests confirm that after almost any dive there would be some asymptomatic bubbles in the body. However, bubble nuclei do not persist for extended periods. Their average halftime is 60 minutes, which means that in 6 hours they would be almost totally eliminated. But it must be remembered that exercise (walking, ladder climbing, surface swims, etc.) constantly produces new bubble nuclei. The dissolved gas will be eliminated until the tissue loading levels reach an equilibrium state with the inspired/alveolar inert gases at ambient pressure (at sea level that would be around 0.76 atm nitrogen and practically zero helium). So there is no "trapped" residual inert gas in the tissues and no chronic inert gas saturation either.

One known practice to break the multiday repetitive cycle is to take one day off every week.

Deep stops revisited

We have seen how theories about deep stops have changed over time. At first, the norm was to follow the schedules generated by raw dissolved-gas models as is. Then some anecdotal data favored inserting arbitrary deep stops by using protocols such as Pyle or GVE. A significant step on the road was tailoring gradient factors as a one-stop M-value reduction mechanism that provides both deep stops and increased conservatism. Finally, automatically generated deep stops became an inherent part of the newer dual-phase models. The vast majority of divers now use one of the latter two.

These emerging trends led the navies around the world to explore further. They are still investing time and money in developing dissolved-gas models such as the USN E-L. They wanted to make sure they would be on the right track if they changed their research directions because there is still a debate on issues such as when to stop, for how long, and how often. More important, is there any evidence that deep stops limit bubble growth or decrease the risk of DCS?

Wayne Gerth, David Doolette, and Keith Gault from NEDU presented their work at DAN's Technical Diving Conference in January 2008.[36] They concluded two distinctive classes of deep stops can be identified. The first class is stops added deeper than the prescribed by a given decompression algorithm. This class would be beneficial under certain circumstances, but it should be remembered that it serves to fix a deficient algorithm. Adding these deep stops cannot allow cutting time off the originally prescribed schedule unless switching to a breathing gas with a higher ppO_2 is associated with the added stops. This is a direct answer to the divers who incorporate a GF low to add deep stops and concurrently increase the GF high above 100%, claiming that bubbles have been suppressed as a result of the introduced deep stops.

The second class is deep stops that arise in comparison of schedules for a given ascent computed with different algorithms and types of supposedly safe ascent criteria. A serious attempt to empirically confirm the theoretical benefits of deep stops by conducting a comparative study was unsuccessful. The results of this attempt were actually at odds with the anecdotal data that favor deep stops.

In June 2008 a workshop was conducted to clarify the role of deep stops and to point out what our current thoughts are as well as indicate future research needs.[37] This workshop brought together a number of high-profile speakers, and some heated discussions took place. Anecdotal data seemed to favor deep stops. On the other hand, two presentations of particular importance from the U.S. Navy and the French Navy indicated they were worse for decompression.

The same work of the NEDU group was presented again. It included the results of a controlled comparative study of two approaches — the traditional dissolved-gas model approach that prescribes schedules that advance rapidly to shallow stops where most of the total stop time is spent, versus the recent dual-phase model approach that prescribes schedules with total stop time skewed toward deeper stops. Divers wearing swimsuits and T-shirts, breathing surface-supplied air via full-face masks, and immersed in 30°C (86°F) water were compressed to 52 meters (170 feet) for 30 minutes. Divers were decompressed with stops prescribed by one of two schedules. The first schedule was generated by the VVal-18 Thalmann dissolved-gas model. The second schedule included deeper stops and was generated by the BVM(3) bubble volume model. The two schedules were matched for total stop time, which was 174 minutes, making the total run time 204 minutes. Unexpectedly, the DCS incidence in the deep-stops schedule were more than three times higher than that in the other schedule. Moreover, two of the DCS cases resulted from following the deep-stops schedule involved rapidly progressing CNS DCS manifestations (Type II DCS, which is more serious than Type I DCS, which usually involves joint pain frequently accompanied by localized swelling, itch and/or skin rash). The study concluded that the deep-stops schedule had a greater risk of DCS than the matched traditional schedule. Slower gas washout or continued gas uptake at the deep stops offset the benefits of reduced bubble growth at deep stops.

The work of Jean-Eric Blatteau, Michel Hugon, and Bernard Gardette from the French Navy was presented, including the results of their studies on deep stops versus conventional schedules for both air and trimix dives until a 100-meter (330-foot) depth. They concluded the advantages of deep stops in human decompression have yet to be demonstrated.

NEDU's original study was released to the public in July 2011, and it struck a nerve with a lot of technical divers. In December 2013 a poll was created on one of the popular technical-diving Internet forums asking whether the participants agreed that NEDU study made sense. A total of 112 technical divers participated, with 91% (102 out of the 112) saying yes, it made sense, and we should actually be more cautious about deep stops. Ten out of the 112 (a little less than 9%) said the study was flawed. The discussion was very interesting.

The participants argued that dual-phase models were unproven, regardless of the large databases of successful dives their advocates collected. On the other hand, empirically derived dissolved-gas models, such as Bühlmann's ZH-L16, worked because they are based on actual decompression trials. In conclusion, a common feeling was that the scientific data we have in hand would not seem to support our current deep-stop practice. It was intriguing to see accomplished technical divers talking about significantly increasing their GF low to the 40 to 50% range to accommodate shallower first stops.

In 2017 a study aiming to compare Bühlmann's ZH-L16 algorithm with gradient factors 30/85 versus the "ratio decompression strategy" was published.[38] Ratio deco is a technique for calculating decompression schedules without using dive tables, decompression software or dive computers. It is generally taught as part of the so-called "Doing It Right" philosophy of diving and the schedules it generates employ deep stops. The study's comparison was based on an analysis of changes in diver circulating inflammatory profiles caused by decompression from a single dive. 51 technical divers performed a single 25-minute trimix dive to 50 meters (165 feet) followed by EAN50 and oxygen for decompression. Twenty-three divers decompressed according to ZH-L16 with gradient factors 30/85 and the rest decompressed according to a ratio deco schedule. Peripheral blood for detection of inflammatory markers was collected before and 90 minutes after diving. Venous gas emboli were measured 30 minutes after diving using 2D echocardiography. Matched groups of 23 recreational divers and 25 swimmers were also enrolled as control groups to assess the effects of decompression from a standard air dive or of exercise alone on the inflammatory profile. Echocardiography observation post dive showed no significant differences between the two decompression procedures. Divers adopting ratio deco showed a worsening of post-dive inflammatory profile compared to the ZH-L16 group. The study concluded that the ratio deco strategy did not confer any benefit in terms of bubbles but showed the disadvantage of increased decompression-associated secretion of inflammatory chemokines involved in the development of vascular damage.

In conclusion, the current position on deep stops was that although they were operational trends, documenting their efficiency was difficult because all the reported cases were uncontrolled anecdotes, not scientific data.

Closed-circuit rebreather (CCR) decompression
A CCR is a breathing apparatus that filters out the carbon dioxide in the exhaled breath. The rest of the exhaled breath, containing some oxygen and some inert gas (probably a valuable amount of helium), gets recycled rather than discharged in the water.

Throughout the dive, the CCR injects oxygen to the breathing loop not only to compensate for the metabolized amount but also to maintain a constant ppO_2 (set point). It also adds bottom mix (diluent) to the loop to hold the ppO_2 from shooting up, especially on the descent.

Let's compare a 20-minute dive to 100 meters (330 feet) using CCR and open circuit (OC). On CCR, I would use trimix (10, 60) as diluent and a set point of 1.2 atm. On OC, I would use trimix (20, 25) as a travel gas and a lean deco mix, trimix (12, 53) as a bottom mix, along with EAN40 and EAN80 for accelerated decompression. Using Bühlmann's ZH-L16B model with gradient factors 30/85, the total run time of the CCR dive is 158 minutes, and the CNS is 73.5%. The OC dive's total run time is 121 minutes, and the CNS is 67.8%. Upon surfacing, the tissue loadings would be as follows:

Cpt #	CCR			OC		
	N$_2$ Load	He Load	Total	N$_2$ Load	He Load	Total
1	1.096m	1.968m	3.064m	3.056m	0.0m	3.056m
2	1.128m	2.101m	3.229m	3.152m	0.0m	3.152m
3	1.204m	2.184m	3.388m	3.66m	0.0m	3.66m
4	1.481m	2.237m	3.718m	4.879m	0.004m	4.883m
5	2.159m	2.31m	4.469m	6.668m	0.072m	6.74m
6	3.159m	2.533m	5.692m	8.335m	0.398m	8.733m
7	4.304m	3.171m	7.475m	9.534m	1.219m	10.753m
8	5.349m	4.353m	9.702m	10.108m	2.44m	12.548m
9	6.149m	5.749m	11.898m	10.161m	3.6m	13.761m
10	6.625m	6.664m	13.289m	9.954m	4.217m	14.171m
11	6.908m	7.053m	13.96m	9.688m	4.406m	14.095m
12	7.104m	7.05m	14.154m	9.395m	4.322m	13.717m
13	7.239m	6.719m	13.957m	9.106m	4.039m	13.145m
14	7.33m	6.149m	13.479m	8.836m	3.631m	12.466m
15	7.39m	5.451m	12.841m	8.598m	3.169m	11.766m
16	7.43m	4.713m	12.143m	8.395m	2.704m	11.099m

Table 8.1: Using Ultimate Planner's "Display tissue loadings upon surfacing" option to compare OC to CCR

We see that for the OC diver, the N$_2$ loadings are always higher than the He loadings for all the compartments. This does not hold true for the CCR diver. That is not unusual because the OC diver switched to EAN40 (0% helium mix) at 30 meters (100 feet), whereas the CCR diver continued breathing a mix containing helium until surfacing. The diluent the CCR diver used is trimix (10, 60), so the ratio of nitrogen to helium is established at 30:60 (1:2) throughout the dive. The nitrogen loadings of the OC diver are always higher than those of the CCR diver, whereas the helium loadings of the CCR diver are always higher than those of the OC diver.

Moreover, although the total run time of the CCR dive is 37 minutes more than that of the OC dive, the total tissue loadings are not always less (higher values are shaded). The good news is that on the surface interval the compartments of the CCR diver will tend to offgas faster than those of the OC diver because the inspired helium content at the surface is practically zero. The helium gradient will be maximized at the surface, and the higher helium loadings on the CCR diver will be eliminated faster than the higher nitrogen loadings on the OC diver. The fact that helium is a fast gas will only decrease its elimination time at the surface.

Concerning ICD, the problem persists on the OC dive. Switching from the bottom mix, which is trimix (12, 53), to trimix (20, 25) on the ascent holds some risk. The problem did not exist on CCR because the CCR diver neither switched gases nor added OC segments to the dive.

If a CCR problem occurs, the diver bails out. In this situation, the diver usually uses the diluent as an OC gas for the deeper part of the ascent before switching to the OC bailout tanks. There is always an urge to end this deeper part of the ascent as soon as possible, as the diluent's gas volume is fairly limited. In this particular situation, a lot of CCR divers prefer using raw Bühlmann ZH-L16 (without deep stops or gradient factors) because it produces shallower stops. To not push the model to the edge of its limits, a safety margin should be added to this "fast bailout" approach. Asymmetric gas kinetics could be used here. This feature is implemented in Ultimate Planner (ZH-L16/U model variation), and it would extend the shallow stops advised by ZH-L16 without introducing deeper ones.

Novel approaches

The results of the most recent studies on the effectiveness of deep stops make me wonder if bubbles really are the underlying cause of DCS. We are trying to control them by introducing deeper stops, but it does not seem to work that way. Maybe they are just an exacerbating factor rather than the causative agent in the progression of DCS. In 2008 a study theorized that the at-depth endothelial dysfunction caused by a temporary loss of haemostasis due to increased total oxidant status is the most significant factor in the progression of DCS.[39] It argues that this increased total oxidant status is a result of breathing oxygen at any pressure because this would cause vasoconstriction. Fortunately, this increase can be prevented by using antioxidants such as Vitamin C. Although the study argues that bubble formation is an "aftermath," it highlights that bubbles have the potential to exacerbate the situation of decompression by damaging the vascular endothelium either through ischemia/reperfusion, physical contact with the endothelium, or by an increase in shear stress. It claims this damage may manifest itself in the form of released endothelial membrane fragments (microparticles).

In 2007 a study on rats revealed the progress of endothelial dysfunction following decompression.[40] Its results verify that bubbles are the causative agents of decompression-induced endothelial damage and bubble amounts are an objective and suitable parameter to predict endothelial dysfunction. It further suggests that levels of endothelial biomarkers postdive may serve as sensitive parameters for assessing bubble load and decompression stress. In 2017 another study on rats suggested that escin, the main active compound in horse chestnut seed extract, had beneficial effects on DCS related to its endothelia-protective properties and thus might be a drug candidate for DCS prevention and treatment.[41]

Some studies suggest a single bout of exercise 24 hours before diving would prevent bubble formation.[42] In 2004, 12 healthy male divers participated in a study where the test subjects underwent a single bout of strenuous exercise for 40 minutes; then they were compressed in a hyperbaric chamber 24 hours later to the equivalent pressure of 18 meters (60 feet) for 80 minutes at a "descent" rate of 10 meters/min (33 feet/min). On their 9-meter/min (30-foot/min) ascent, they stopped at 3 meters (10 feet) for 7 minutes. After reaching the surface pressure, an ultrasonic scanner monitored venous gas bubbles. The results demonstrated that compared with dives without preceding exercise, the average number of bubbles in the pulmonary artery was significantly reduced (from 0.98

to 0.22 bubbles per square centimeter). By undergoing predive exercise, the maximum bubble grade was decreased from 3 to 1.5, thus safety was increased.

A study on rats demonstrated that the administration of a nitric-oxide releasing agent 30 minutes prior to diving has the same effect of predive exercise.[43] This confirmed the results of a former study suggesting that nitric oxide synthase inhibition increases bubble formation.[44] No tests have been conducted on humans so far.

Nitrogen and helium kinetics – back to basics

We've seen in Chapter 5 that helium is almost 2.65 times "faster" than nitrogen. We use this value, which is called the diffusivity ratio, to obtain the helium compartment halftimes in relation to those of nitrogen.

In December 2014, a study on sheep found that altering blood flow did not reveal differences between nitrogen and helium kinetics in brain or in skeletal miracle.[45] It bothered a lot of technical divers, since it concluded that it was inappropriate to assign substantially different time constants for nitrogen and helium in all compartments in decompression algorithms. The published results showed that in the brain, there were no differences in nitrogen and helium kinetics, while hind-limb models indicated nitrogen kinetics were slightly faster than helium. It illustrated that in both the brain and the hind limb, the blood-tissue exchange of nitrogen was similar to that of helium.

These findings were consistent with the results of another study published in April 2015.[46] As it was widely believed that trimix bounce dives could be conducted with substantially reduced decompression times than corresponding heliox dives, decompression efficiency was assessed by comparing the incidence of DCS following decompression dives using MK 16 MOD 1 UBAs (1.3 atm set point) with either heliox (88% He / 12% O_2) or trimix (44% He / 44% N_2 / 12% O_2) diluent. Both trimix and heliox dives followed the identical depth/time schedule (60 meters [200 feet] for 40 minutes bottom time followed by 119 minutes of decompression stops). This schedule was selected for having the largest difference in estimated probabilities of DCS between trimix and heliox among a range of candidate schedules that were practicable to man-test and operationally relevant. Fifty man-dives were completed on the heliox schedule with no diagnosed incidents of DCS, whereas 46 man-dives were completed on the trimix schedule with two diagnosed incidents of DCS. It concluded that decompression from trimix bounce dives was not more efficient than decompression from heliox bounce dives.

Summary

This chapter concluded our look at the decompression theory by exploring several topics. In this chapter, we've explored a grab bag of additional points of interest to divers. Along the way, we answered some of the questions raised on Internet forums, such as the possibility of getting bent because of using oxygen, the effect of dehydration, the influence of temperature variation, the impact of exercise, planning dives at altitude, adding conservatism, the merits of ultra long halftimes, the side effects of suffering a PFO, and the viability of washout treatment.

We've also seen what accomplished technical divers do when they don't have enough pre-flight surface intervals, and based on the existing decompression models, we proposed a formal method for accelerating the no-fly time. Some decompression principles were also discussed, including the oxygen window, asymmetric gas kinetics, CCR decompression, along with a second look at the deep stops. Controversial issues include multilevel diving, omitted decompression, in-water recompression, acclimatization, and some novel approaches to mitigate the risk of DCS.

Coda

I hope this book illustrates some of the recent developments in decompression theory. I linked the references to the text by using superscript numbers to make it easy for you to explore further. As we have seen, over a relatively short period the trends can change 180 degrees: *End of bottom time, get out of the deep! Oh no, that is why you are not feeling well postdive. . . . Stop "somewhere" while ascending, here is a protocol. . . . Come on folks, the entire methodology is flawed. . . . Let's beat the bubbles rather than treating them. . . . This new approach advises deep stops without adding any fudge factors. . . . And by the way, multiday diving increases the risk of DCS. . . . We should "reduce the gradient" to account for that. Then years later, it is suggested to not reduce anything. . . . It seems multiday diving is good for acclimatization. . . . And we are not sure about this deep-stop thing anymore. . . . Actually bubble formation might just be an aftermath, not the causative agent in the progression of DCS.*

Even basic things such as the uptake and elimination of inert gas remain only partially understood today. Only some gross measurements have been made, and the results were exponential. Of course when it comes to exact bubble production mechanisms and the interaction of pressures and temperatures, things become much more complicated.

Now that machine learning (a subdomain of artificial intelligence) is becoming more prevalent, what I'd like to see as a future research trend is feeding the computer with the data obtained from dive profiles including the diver's gender, age, body mass index (BMI), health condition, etc., then training the computer with these data to get "tailored" deco models. For instance, a 47-year-old male diver with BMI 24 and no known health conditions might ask for a tailored, 0.1% accepted risk deco schedule for a 25-minute CCR dive to 100 meter (330 foot) depth using trimix (10, 60) and a set point of 1.2 atm in salt water at sea level where the water temperature ranges from 25° to 27°C (77° to 81°F) and the thermal protection is a 7mm wetsuit. One good thing about this approach is the possibility of continuous model training—with every executed dive the models "accuracy" could be increased.

Now that you know what it means that decompression is still far from being an exact science, can you answer the question in the introduction of this book? Concentrating only on the decompression part and discarding every other aspect of the dive I referred to, should I have completed the deco schedule my computer dictated?

APPENDIX A
Using Hydrogen as a Diving Gas

One of the main disadvantages of using heliox (helium-oxygen mix) in deep commercial diving is the high-pressure nervous syndrome (HPNS) associated with breathing helium at elevated pressures. Adding nitrogen to the heliox mix (making it trimix) helps ameliorate the HPNS, but using trimix at extreme depths can lead to narcosis, which is why hydrogen was considered as a replacement. In 1988 Comex commercial divers reached a maximum depth of 534 meters (1,750 feet) at sea as part of their epic project Hydra VIII. Their diver Theo Mavrostomos went 701 meters (2,230 feet) in the hyperbaric chamber during the Hydra X project in 1992.

Since then he remains "the deepest man in the world."[1] Studies conducted by Comex confirmed that, at the right pressure, hydrogen can reduce HPNS.[2]

In recent years the availability of helium has decreased, and its cost has significantly increased, leading some technical divers to frequently ask about the possibility of switching to the second-best nitrogen substitute — hydrogen — for sports diving.

Hydrogen is inexpensive. It is more soluble in lipid than helium, so it is more narcotic than helium yet still less narcotic than nitrogen. According to E.B. Smith, theoretically speaking, the narcotic potency of hydrogen should be similar to that of a mixture containing 50% nitrogen and 50% helium.[3]

Comex studied the various narcotic effects and concluded the narcotic effect of hydrogen is not the same as nitrogen. As per their studies, nitrogen narcosis is comparable to alcoholic intoxication, whereas hydrogen narcosis results in symptoms that seem closer to those produced by hallucinogenic drugs such as LSD and mescaline. Based on their practical studies, Comex estimated the narcotic potency of hydrogen to be around ¼ that of nitrogen.[4]

Hydrogen's thermal conductivity is quite high, higher than that of nitrogen and even higher than that of helium. It is a colder gas to breathe, which indicates heat loss would occur more rapidly while breathing hydrogen-based mixtures. The speed of sound in hydrogen is even more than that in helium. A diver breathing a hydrogen-based mix would expect his voice to sound unusually high-pitched.

Hydrogen is lighter than helium so it is better in terms of breathing resistance at depth. Its lightness also makes it faster in terms of saturation and desaturation. The fundamental drawback associated with using hydrogen is that it becomes flammable and possibly explosive when mixed with oxygen percentages over 4%.

Saturation and desaturation speeds of inert gases in body tissues are inversely proportional to the square root of their molecular weights. By applying this to both helium and hydrogen in comparison with nitrogen, we get the following:

Speed (He) / Speed (N₂) = SQRT (28.01) / SQRT (4.00) = 2.65 approximately

Speed (H₂) / Speed (N₂) = SQRT (28.01) / SQRT (2.02) = 3.72 approximately

This means helium is almost 2.65 times faster than nitrogen, whereas hydrogen is some 3.72 times faster than nitrogen. These are simply the diffusivity ratios, and we use them to obtain the compartment halftimes in relation to those of nitrogen.

Let's take a 1-hour compartment (halftime = 60 minutes) as an example. This hour is the time it takes the compartment to get half-filled (saturated) or half-emptied (desaturated), given that the gas in action is nitrogen. For the very same compartment, its helium halftime is 60 / 2.65 = 22.64 minutes approximately. For hydrogen it's 60 / 3.72 = 16.13 minutes approximately.

As the size of bubbles depends on the amount of dissolved gas, the higher the gas' solubility in the blood, the bigger the bubbles it promotes. The solubility in blood of nitrogen, helium, and hydrogen are 0.0122, 0.0087 and 0.0149 atm[-(1)] respectively. So we would expect the hydrogen bubbles to be bigger than both the helium and nitrogen bubbles.

Concerning ICD, the models that aim to predict ICD are solubility-only models. As per our current understanding, the effect of diffusivity is negligible. Based on that, theoretically speaking ICD should be less of a problem with hydrogen-based mixtures if applying the rule of fifth. On the practical side, as you cannot increase the oxygen percentage in the hydrogen-based mixtures, all you can do is gradually reduce the relative concentration of hydrogen (i.e., substitute some of the hydrogen content with nitrogen while keeping the oxygen content constant).

For the sake of demonstration, let's plan a dive using hydrox (hydrogen-oxygen mix). I have compiled a special version of Ultimate Planner to work with hydrogen instead of helium. To use VPM-B, I needed a set of halftimes corresponding to that of nitrogen. This was obtained by dividing the nitrogen halftimes by 3.72. I also needed an *initial critical radius* for hydrogen. (Consult Chapter 6 for more information about the critical radii.) The default initial critical radii values that we feed the VPM-B algorithm are 0.55 and 0.45 micron for nitrogen and helium respectively. These values were set so that a fully saturated diver would have allowable supersaturation pressure that was consistent with the experimentally determined maximum allowable saturation depth for direct ascent to the surface. Now that I do not have the facilities to do such experiments on hydrogen, let's try another approach.

One of the key concepts Bühlmann demonstrated is the tolerated partial pressures of two different gases in the same compartment will vary according to their solubility coefficients in the *transport medium* that delivered those gases to that compartment. The transport medium in our case is the blood plasma, so we will use the solubility in blood of nitrogen, helium, and hydrogen along with the initial critical radii of nitrogen and helium to extrapolate a value for the initial critical radius of hydrogen. A simple linear extrapolation results in approximately 0.627 micron. As we already know, this means VPM-B will not allow any hydrogen bubbles with *smaller* radii than this critical value to grow, which is consistent with our expectation that the hydrogen bubbles would be bigger than both the helium and nitrogen bubbles.

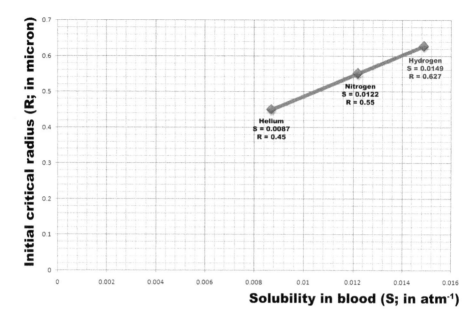

Extrapolation to get the initial critical radius of hydrogen

We'll plan a 20-minute dive to a 100-meter (330-foot) depth. The descent rate is 20 meters/min (66 feet/min), so the actual time we'll spend at 100 meters (330 feet) is 15 minutes. Since the maximum allowable content of oxygen in the hydrox mix is 4%, this mix could be used only as a bottom mix. It is not breathable at the surface. To support life, we need a ppO_2 of at least 0.16 bar, so we will use air as a travel gas to 40 meters (130 feet) then switch to a flushing gas to remove excess oxygen from the diver's body. This flushing gas consists of 4% O_2 and 96% N_2, and we'll use it for three minutes to give the blood sufficient time to circulate around the entire body. To combat narcosis, we cannot descend with the flushing gas, so these 3 minutes will be at 40 meters (130 feet). This makes the actual time we will spend at 100 meters (330 feet) 12 minutes. Eventually we will switch to hydrox (4% O_2, 96% H_2) at 40 meters (130 feet), continue descending, then spend the rest of our bottom time. The ascent rate is 9 meters/min (30 feet/min). On our ascent, we will switch back to the flushing gas at 66 meters (220 feet), breathe it for 3 minutes while doing our decompression stops, then we will start using air. As per our current understanding, oxygen is narcotic, probably equally narcotic to nitrogen, so our flushing gas will still have the same END of air. At 30 meters (100 feet) we will switch to EAN40, and our final switch will be at 9 meters (30 feet) to EAN80. So the plan is as follows:

- Descend to 40m (130ft) on air in 2 minutes.

- Switch to flushing gas (4% O_2, 96% N_2) at constant depth for 3 minutes.

- Switch to hydrox (4% O_2, 96% H_2) and descend to 100m (330ft) in 3 minutes.

- Remain at a constant depth at 100m (330ft) for 12 minutes.

- Switch to flushing gas on the ascent at 66m (220ft) for 3 minutes.

- Switch to air after 3 minutes of breathing the flushing gas (at 57m, 190ft).

- Switch to EAN40 at 30m (100ft).

- Switch to EAN80 at 9m (30ft).

We will use VPM-B with 12% conservatism (your dive computer might call that VPM-B +2). The total run time of this dive is 270 minutes. The decompression stops start at 96 meters (320 feet), and the total CNS is 119.9%.

For the same dive on trimix, I would use trimix (20, 25) as travel gas to 60 meters (200 feet) then trimix (12, 53) as bottom mix. I would switch back to the travel gas on my ascent at 69 meters (230 feet). I will not change the EAN40 and the EAN80. The total run time of this dive is 114 minutes. The decompression stops start at 72 meters (240 feet), and the total CNS is 63.5%.

We know commercial diving operations have tried hydrogen-helium-oxygen mixes, which they frequently call hydreliox and less frequently hydroheliox. Since we want to substitute helium for its cost and increasing scarcity, now the question is if we can add nitrogen to the hydrox mix. Has anyone tried this before?

In 1944 Arne Zetterström (1917-1945) of the Swedish Navy conducted 2 open-ocean dives to 40 and 70 meters (130 and 230 feet), breathing a 3-gas mixture containing 72% hydrogen, 24% nitrogen and 4% oxygen.[5] On August 7, 1945, he lost his life on a 160-meter (525-foot) dive when one of the winch operators raised him directly to the surface without doing his decompression stops. His death had nothing to do with him breathing the hydrogen-nitrogen-oxygen one year earlier.

Now let's reschedule the very same dive with our new bottom mix. It contains 63% hydrogen, 33% nitrogen and 4% oxygen. According to Comex estimation on hydrogen narcosis potency, the narcotic potency of this mix would be equivalent to that of trimix (4, 47). The total run time of this dive is 137 minutes. The decompression stops start at 90 meters (300 feet), and the total CNS is 58.8%. Switching from this hydrogen-nitrogen-oxygen bottom mix to the flushing gas on ascent supposedly implies less risk of getting an ICD hit than switching from hydrox to the flushing gas.

It is worth noting that the 4% oxygen limit prohibits using hydrogen with CCR. Injecting fresh oxygen to compensate for metabolic oxygen consumption would result in an explosion.

One of the most innovative approaches for decreasing the DCS risk is biochemical decompression. The concept is increasing the tissues' inert gas washout rate by injecting metabolizing microbes into the large intestine. This approach was tested with success in pigs during simulated hydrogen dives.[6] When a probabilistic model to predict DCS outcome in pigs during hydrogen dives was used, the model suggested that animals injected with hydrogen metabolizing microbes have a decreased probability of DCS.[7] Application to man is awaiting additional studies.

Calculating the Acceleration in Post-Diving No-Fly Time Associated with Breathing Surface Oxygen

Abstract

BACKGROUND: Some experienced scuba divers breathe oxygen on the surface after the last dive to accelerate their pre-flight surface interval.[1] This study investigates the possibility of applying a computational algorithm to estimate the time gain associated with this practice without attempting to calculate the no-fly time itself. It also compares the no-fly time savings when helium is added to the bottom mix, and when the timing of applying the oxygen on the surface is changed. METHODS: A stand-alone computer program to plan dives was developed and a recursive algorithm was implemented to calculate the no-fly time savings. A 40 minute open circuit dive to a maximum depth of 55 meters (180 feet) was simulated, first using air as the bottom mix then using Trimix21/25 (21% oxygen, 25% helium). The decompression gases were Nitrox40 (40% oxygen, 60% nitrogen) starting at 30 meters (100 feet) then Nitrox80 (80% oxygen, 20% nitrogen) starting at 9 meters (30 feet). RESULTS AND CONCLUSIONS: Estimating the no-fly time acceleration associated with breathing oxygen on the surface is possible without calculating the no-fly time itself. The gain is lower in case the bottom mix contains a percentage of helium. The gain increases with the delay of oxygen application.

Introduction

A flying-after-diving workshop was hosted at Divers Alert Network in May 2, 2002.[2] The objective was to review the state of knowledge of flying after diving and to discuss the need for new guidelines for recreational diving.

The workshop recommendations were as follows:[3]

• For a single no-decompression dive, a minimum pre-flight surface interval (SI) of 12 hours is suggested.

• For multiple dives per day or multiple days of diving, a minimum pre-flight SI of 18 hours is suggested.

• For dives requiring decompression stops, there is little experimental or published evidence on which to base a recommendation. A pre-flight SI substantially longer than 18 hours appears prudent.

All recommendations assume air dives followed by flights at cabin altitudes of 610 to 2,438 meters (2,000 to 8,000 feet) for divers who do not have symptoms of decompression sickness (DCS). These recommendations were not predicted by mathematical models, but rather suggested from empirical data based on observations and actual occurrences of DCS.

Divers doing dives that require decompression stops usually subscribe to a minimum pre-flight SI of 24 hours. Some experienced technical divers breathe oxygen on the surface after the last dive to accelerate their pre-flight SI. However, the majority of them don't seem to have solid, established protocols.[4] Developing a tool to estimate the time gain associated with this practice appears shrewd.

Methods

A stand-alone computer program incorporating VPM-B was developed to plan dives. The program uses the 16 halftime compartment set of Albert Bühlmann's ZH-L16 model.[5] A special algorithm was implemented to calculate the total inert gas loadings in the array of tissue compartments directly after the dives. The algorithm then re-calculates the total tissue loadings after breathing oxygen on the surface for a determined time period (T1) and loops back on all the compartments to calculate how much time would have been needed to reach the same total loading without breathing oxygen (T2). As soon as any compartment reaches the same total loading the program exits the loop. The gain is T2 minus T1. The accuracy of the calculations is 1 minute.

The calculation procedure is a recursive application of two central equations advised by John Scott Haldane.[6] They are employed in popular decompression models like VPM-B and ZH-L16 to calculate the uptake and elimination of inert gases during intervals at constant depth, both in and out of the water.

$$p - pa = (p - pa) \exp (-k * t) \qquad (1)$$

p: inert gas partial pressure
pa: inert gas ambient partial pressure
k: inert gas time constant
t: time

$$k = \ln (2) / h \qquad (2)$$

h: inert gas compartment halftime

A 40-minute open-circuit dive to a maximum depth of 55 meters (180 feet) was simulated. The descent rate was 20 m/min (66 ft/min) and the ascent rate was 10 m/min (33 ft/min) up till the 6 meter (20 foot) mark then was slowed down to 3 m/min (10 ft/min). The decompression gases were Nitrox40 (40% oxygen, 60% nitrogen) starting at 30 meters (100 feet) then Nitrox80 (80% oxygen, 20% nitrogen) starting at 9 meters (30 feet). The decompression schedules were generated assuming the bottom mix is air then re-generated assuming the bottom mix is Trimix21/25 (21% oxygen, 25% helium). A 12% conservatism level was applied to both dives.

Results

Bottom mix: air
VPM-B conservatism: 12%
Total run time: 107 minutes
No-fly time saving
(breathing pure oxygen for 60 minutes, directly after surfacing): 116 minutes

No-fly time saving
(breathing pure oxygen for 60 minutes, 60 minutes after surfacing): 173 minutes
No-fly time saving
(breathing pure oxygen for 60 minutes, 120 minutes after surfacing): 230 minutes
No-fly time saving
(breathing pure oxygen for 60 minutes, 180 minutes after surfacing): 288 minutes
No-fly time saving
(breathing pure oxygen for 60 minutes, 240 minutes after surfacing): 347 minutes

Bottom mix: Trimix21/25
VPM-B conservatism: 12%
Total run time: 96 minutes
No-fly time saving
(breathing pure oxygen for 60 minutes, directly after surfacing): 73 minutes
No-fly time saving
(breathing pure oxygen for 60 minutes, 60 minutes after surfacing): 110 minutes
No-fly time saving
(breathing pure oxygen for 60 minutes, 120 minutes after surfacing): 147 minutes
No-fly time saving
(breathing pure oxygen for 60 minutes, 180 minutes after surfacing): 183 minutes
No-fly time saving
(breathing pure oxygen for 60 minutes, 240 minutes after surfacing): 218 minutes

Discussion

The No-Fly Time Accelerator (NFTA) is an attempt at developing a usable tool for quantifying the gain of utilizing an already-employed strategy.

Trying to calculate the no-fly time puts the developer on the horns of a dilemma. Using the commercial cabin altitude ranges result in much less pre-flight SI than the recommended. Increasing the cabin altitude is an ad-hoc and would result in a calibration problem with the calculated time savings.

For these reasons, NFTA neither aims nor tries to calculate the no-fly time. It does not use the flight or altitude data either. The only input that NFTA receives from the decompression model is tissue loadings upon surfacing, so it is irrelevant to NFTA whether a dual phase model like VPM-B or a dissolved gas model like ZH-L16 is used to plan the dives.

NFTA employs the straight forward approach of calculating how much faster the tissue compartments de-saturate on the surface when breathing oxygen instead of air for a certain period of time. The equations used are already employed in scores of decompression planning tools incorporating popular decompression models like ZH-L16 and VPM-B.

NFTA does not assume that a particular compartment, or a set of compartments, control the no-fly time. The recursive loop does not look at specific compartments. It stops as soon as any compartment fulfills the condition: total tissue loading after breathing oxygen for duration T1 = total tissue loading after breathing air for duration T2. That's why the time savings are presumably conservative.

The gain is lower in case the bottom mix contains a percentage of helium. This is expected, as the inspired helium content on the surface is negligible anyway, so breathing oxygen won't accelerate helium elimination.

The gain increases with the delay of oxygen application. Although this might seem at odds with the current understanding of offgassing gradients, it's a normal mathematical result for employing exponential equations. To illustrate, let's say (just for demonstration) that the diver surfaced with the 5 minute halftime compartment holding a nitrogen load of 2.76 bar. If the diver breathes pure oxygen for 5 minutes directly after surfacing, the tissue tension will be ((2.76 – 0.0) / 2) + 0.0 = 1.38 bar. After 5 more minutes but on air, the tissue tension will be ((1.38 – 0.76 [alveolar ppN2]) / 2) + 0.76 = 1.07 bar. On the other hand, if the diver delayed the surface oxygen for 5 minutes, the tissue tension would be ((2.76 – 0.76) / 2) + 0.76 = 1.76 bar just before the oxygen application, and ((1.76 – 0.0) / 2) + 0.0 = 0.88 bar after breathing oxygen for 5 minutes. So delaying the surface oxygen indeed boosted offgassing.

Conclusions

Estimating the no-fly time acceleration associated with breathing oxygen on the surface is possible without calculating the no-fly time itself. The gain is lower in case the bottom mix contains a percentage of helium. The gain increases with the delay of oxygen application.

ACKNOWLEDGMENTS

This book contains several interviews with researchers, accomplished divers, industry professionals, and software developers. I am grateful to all participants. Bret Gilliam in particular provided some of the most interesting and valuable accounts. Dr. Michael Powell, a retired NASA researcher who is best known as Dr. Deco, allowed me to quote some of his precious writings. Without his help this book would not have been as diverse as it is. Powell has a question-and-answer forum on ScubaBoard.com called "Ask Dr. Decompression" that might be of interest to readers.

I am particularly grateful to Jurij Zelič, the developer of VPM Open and VPM Mixer. When I started the no-fly time acceleration research, he was the man behind the scenes. The assistance in software development I received from him was invaluable.

I would also like to thank T. Timothy Smith for the cover photo of the first edition and René Andersen for the cover photo of the second edition.

REFERENCES

Chapter 1: Historical Perspective

1. Powell MR. Liquids as a hole: nucleation in diving. *Tech Diving Mag*, Issue 5, December 2011.

2. Stillson GD. *Report on Deep Diving Tests*. U.S. Navy Bureau of Construction and Repair. Washington, DC: Government Printing Office; 1915.

3. Hempelman HV. *Investigation into the Decompression Tables. Report III, Part A. A new theoretical basis for the calculation of decompression tables.* Royal Naval Personnel Research Committee, Underwater Physiology Subcommittee Report 131. London: Medical Research Council; 1952.

4. Hempleman HV. *The Unequal Rates of Uptake and Elimination of Tissue Nitrogen Gas in Diving Procedures.* Royal Naval Personnel Research Committee, RNP 62/1019, UPS 195, RNPL 5/60. London: Medical Research Council; 1960.

Chapter 2: Basic Decompression Principles

1. British Admiralty. *Report of a Committee Appointed by the Lords Commissioners of the Admiralty to Consider and Report Upon the Conditions of Deep-Water Diving.* London: H.M. Stationery Office; 1907.

2. Boycott AE, Damant GCC, Haldane JS. The prevention of compressed air illness. *J Hyg* (London). Jun 1908; 8(3): 342–443.

3. Hamilton RW, Rogers RE, Powell MR, Vann RD. *The DSAT Recreational Dive Planner: Development and validation of no-stop decompression procedures for recreational diving.* Tarrytown, NY: Diving Science and Technology Inc. and Hamilton Research Ltd; 1994.

4. U.S. Department of the Navy. *U.S. Navy Diving Manual, Revision 4.* NAVSEA 0994-LP-100-3199. SS521-Ag-PRO-010. Washington, DC: Naval Sea Systems Command; 2001.

Chapter 3: Dissolved-Gas (Haldanean) Models

1. Boycott AE, Damant GCC, Haldane JS. The prevention of compressed air illness. *J Hyg* (London). Jun 1908; 8(3): 342–443.

2. Workman RD. *Calculation of Air Saturation Decompression Tables.* Washington, D.C.: U.S. Navy Experimental Diving Unit; 1957.

3. Workman RD. *Calculation of Decompression Schedules for Nitrogen-Oxygen and Helium-Oxygen Dives.* Research Report 6-65. Washington, D.C.: U.S. Navy Experimental Diving Unit; 1965.

4. Schreiner HR, Kelley PL. A pragmatic view of decompression. In: Lambertsen CJ, ed. *Proceedings of the 4th Symposium on Underwater Physiology.* New York: Academic Press; 1971:205-219.

5.   Bühlmann AA. *Tauchmedizin*. Berlin: Springer-Verlag; 1995.

6.   Pyle R. The importance of deep safety stops: rethinking ascent patterns from decompression dives. *DeepTech* 1995; 5:64.

7.   Marroni A, Bennett PB, Cronjé FJ, Cali-Corleo R, Germonpre P, Pieri M, Bonuccelli C, Balestra C. A deep stop during decompression from 82 fsw (25 msw) significantly reduces bubbles and fast tissue gas tensions. *UHM* 2004; 31(2).

Chapter 4: Nitrox

1.   Gilliam B. Behind the 1990s controversy over technical diving. *Tech Diving Mag*, Issue 2, March 2011.

2.   Lafère P, Balestra C, Hemelryck W, Donda N, Sakr A, Taher A, Marroni S, Germonpré P. Evaluation of critical flicker fusion frequency and perceived fatigue in divers after air and enriched air nitrox diving. *Diving Hyperb Med.* 2010 Sep; 40(3):114-8.

3.   Marinovic J, Ljubkovic M, Breskovic T, Gunjaca G, Obad A, Modun D, Bilopavlovic N, Tsikas D, Dujic Z. Effects of successive air and nitrox dives on human vascular function. *Eur J Appl Physiol.* 2012 Jun; 112(6):2131-7.

4.   Joyner JT, ed. *NOAA Diving Manual: Diving for Science and Technology, 4th ed.* Flagstaff, AZ: Best Publishing with U.S. Department of Commerce, National Technical Information Service; 2001: 3.1-3.36.

Chapter 5: Mixed Gas

1.   Falcon (Salvage Ship), Behnke AR. *Log of Diving During Rescue and Salvage Operations of the USS Squalus: Diving Log of USS Falcon, 24 May 1939-12 September 1939.* Kensington, Md.: Reprinted by Undersea and Hyperbaric Medical Society; 2001.

2.   Gilliam B. A practical discussion of nitrogen narcosis. *Tech Diving Mag*, Issue 2, March 2011.

3.   Miller KW, Paton WD, Smith RA, Smith EB. The pressure reversal of general anesthesia and the critical volume hypothesis. *Molec Pharmacol* 1973; 9:131-143.

4.   Gilliam B. Zacaton: the tragic death of Sheck Exley. *Tech Diving Mag*, Issue 12, September 2013.

5.   Hesser CM, Fagraeus L, Adolfson J. Roles of nitrogen, oxygen, and carbon dioxide in compressed-air narcosis. *Undersea Biomed Res.* 1978 Dec; 5(4): 391-400. PMID: 734806.

6.   Joyner JT, ed. *NOAA Diving Manual: Diving for Science and Technology, 4th ed.* Flagstaff, AZ: Best Publishing with U.S. Department of Commerce, National Technical Information Service; 2001: 16.1-16.15.

7. Keller HA. Use of multiple inert gas mixtures in deep diving. In: Lambertsen CJ, ed. *Proceedings of the 3rd Symposium on Underwater Physiology*. Baltimore: Williams and Wilkins; 1967: 267-274.

8. Bühlmann AA. Decompression theory: Swiss practice. In: Bennett PB, Elliott DH, eds. *The Physiology and Medicine of Diving and Compressed Air Work, 2nd Edition*. London: Baillière Tindall; 1975.

9. Shilling CW, Werts MF, Schandelmeier NR, eds. *Underwater Handbook*. New York: Plenum Press; 1976.

10. Powell MR. Liquids as a hole: nucleation in diving. *Tech Diving Mag*, Issue 5, December 2011.

11. Mitchell SJ, Doolette DJ. Pathophysiology of inner ear decompression sickness: potential role of the persistent patent foramen ovale. *Diving Hyperb Med*. 2015;45:105-110.

12. Burton S. Isobaric counterdiffusion. *Scuba Engineer*. December 2004, Rev: 2011.

13. Doolette DJ, Mitchell SJ. Biophysical basis for inner-ear decompression sickness. *J Appl Physiol*. 2003; 94:2145-2150. doi:10.1152/japplphysiol.01090.2002.

Chapter 6: Dual-Phase (Bubble) Models

1. Powell MR. Diving amongst the stars: NASA and its contributions to recreational scuba. *Tech Diving Mag*, Issue 3, June 2011.

2. Wilmshurst P. Brain damage in divers. *BMJ*. 1997; 314:689-90.

3. Erdem I, Yildiz S, Uzun G, Sonmez G, Senol MG, Mutluoglu M, Multu H, Oner B. Cerebral white-matter lesions in asymptomatic military divers. *Aviat. Space Environ. Med*. 80(1). doi: 10.3357/ASEM.2234.2009.

4. Powell MR. Liquids as a hole: nucleation in diving. *Tech Diving Mag*, Issue 5, December 2011.

5. Hills BA. A fundamental approach to the prevention of decompression sickness. *SPUMS J*. 1978; 8(2):20-47.

6. Behnke AR. The application of measurements of nitrogen elimination to the problem of decompressing divers. *US Nav Med Bull*. 1937; 35: 219–240.

7. Powell MR. Diving amongst the stars: NASA and its contributions to recreational scuba. *Tech Diving Mag*, Issue 3, June 2011.

8. Maiken E. VPM supersaturation gradients. Available at: http://www.decompression.org/maiken/VPM/RDPW/VPMech4/VPMech4.htm

9. Hennessy TR, Hempleman HV. An examination of the critical released gas volume concept in decompression sickness. *Proc. R. Soc. Lond. B*. 1977; 197:299-313.

10. Yount DE, Hoffman DC. On the use of a bubble formation model to calculate diving tables. *Aviat. Space Environ. Med.* 1986; 57:149-56.

11. Salama A. VPM-B variations: /E, /GFS and /U. *Tech Diving Mag*, Issue 5, December 2011.

12. Lang MA, Lehner CE, eds. *Proceedings of Reverse Dive Profiles Workshop.* October 29-30, 1999. Smithsonian Institution, Washington, DC.

13. Wienke BR, O'Leary TR. Reduced gradient bubble model: diving algorithm, basis, and comparisons. Available at: http://www.scuba-doc.com/rgbmim.pdf

Chapter 7: Other Decompression Models

1. Hempelman HV. *Investigation into the Decompression Tables. Report III, part A. A new theoretical basis for the calculation of decompression tables.* Royal Naval Personnel Research Committee, Underwater Physiology Subcommittee Report 131. London: Medical Research Council; 1952.

2. Hempleman HV. *The Unequal Rates of Uptake and Elimination of Tissue Nitrogen Gas in Diving Procedures.* Royal Naval Personnel Research Committee, RNP 62/1019, UPS 195, RNPL 5/60. London: Medical Research Council; 1960.

3. Kidd D, Stubbs RA, Weaver RS. Comparative approaches to prophylactic decompression. In: Lambertsen CJ, ed. *Proceedings of the 4th Symposium on Underwater Physiology.* New York: Academic Press; 1971:167–177.

4. Nishi RY, Lauckner GR. *Development of the DCIEM 1983 Decompression Model for Compressed Air Diving.* Downsview, Ontario: Defence and Civil Institute of Environmental Medicine; 1984.

5. Thalmann ED. Phase II Testing of *Decompression Algorithms for use in the U.S. Navy Underwater Decompression Computer.* Technical Report 1-84. Panama City, FL: Navy Experimental Diving Unit; 1984.

6. Doolette DJ, Gault KA, Gerth WA, Murphy FG. U.S. Navy Dive Computer Validation. In: S. Lesley Blogg, Michael A. Lang, Andreas Møllerløkken, eds. *Proceedings of Validation of Dive Computers Workshop*, at the 37th Annual Meeting of the European Underwater and Baromedical Society, Gdansk, Poland. August 24, 2011.

7. Gerth WA, Doolette DJ. *VVal-18 and VVal-18M Thalmann Algorithm Air Decompression Tables and Procedures.* TA 01-07. NEDU TR 07-09. Panama City, FL: Navy Experimental Diving Unit; 2007.

8. Parker EC, Survanshi SS, Massell PB, Weathersby PK. Probabilistic models of the role of oxygen in human decompression sickness. *J Appl Physiol.* 1998 Mar; 84(3):1096-102.

9. Thalmann ED, Kelleher PC, Survanshi SS, Parker EC, Weathersby PK. *Statistically Based Decompression Tables XI: Manned Validation of the LE*

Probabilistic Model for Air and Nitrogen-Oxygen Diving. Technical Report May 1991-May 1993. Bethesda, Md.: Naval Medical Research Center; 1999.

10. Walker JR, Hobbs GW, Gault KA, Leong ML, Howle LE, Freiberger JJ. Decompression risk analysis comparing oxygen and 50% nitrox decompression stops. 2010 Undersea and Hyperbaric Medical Society Annual Scientific Meeting. *Undersea Hyperb Med.* 2010; 37(4).

11. Nikolaev VP. Probabilistic model of decompression sickness based on stochastic models of bubbling in tissues. *Aviat Space Environ Med.* 2004 Jul; 75(7):603-10.

12. Besnard S, Philippot M, Hervé PH, Porcher M, Arbeille PH. Intravascular ultrasound contrast agent particles in the cerebral, renal and lower limb arteries — consequence on diving physiology. In: Germonpré P, Balestra C, eds. *Proceedings of the 28th Annual Scientific Meeting of the European Underwater and Baromedical Society*, Bruges, Belgium, 4-8 September 2002.

13. Imbert JP, Paris D, Hugon J. The arterial bubble model for decompression tables calculations. In: Grandjean B, Meliet J, eds. *Proceedings of the 30th Annual Scientific Meeting of the European Underwater and Baromedical Society*, Corsica, France, September 15-19, 2004.

14. Brubakk AO, Arntzen AJ, Wienke BR, Koteng S. Decompression profile and bubble formation after dives with surface decompression: experimental support for a dual phase model of decompression. *Undersea Hyperb Med.* 2003; 30(3):181-193. PMID: 14620098.

15. Brubakk AO, Gutvik C. Optimal decompression from 90 msw. *Proceedings of Advanced Scientific Diving Workshop*, February 23-24, 2006. Smithsonian Institution, Washington, DC.

16. Goldman S. A new class of biophysical models for predicting the probability of decompression sickness in scuba diving. *J Appl Physiol.* 2007; 103(2):484-493.

17. Howle LE, Weber PW, Hada EA, Vann RD, Denoble PJ. The probability and severity of decompression sickness. *PLoS ONE* 12(3):e0172665. https://doi.org/10.1371/journal.pone.0172665.

Chapter 8: Various Topics

1. Behnke AR. The isobaric (oxygen window) principle of decompression. In: *The New Thrust Seaward.* Transactions of the Third Annual Conference of the Marine Technology Society, 5-7 June, San Diego. Washington DC: Marine Technology Society; 1967.

2. Brian JE Jr. Gas exchange, partial pressure gradients, and the oxygen window. Department of Anesthesia, University of Iowa College of Medicine; 2001.

3. Van Liew HD, Conkin J, Burkard ME. The oxygen window and decompression bubbles: estimates and significance. *Aviat Space Environ Med.* 1993 Sep; 64(9 Pt 1):859-65.

4. Powell MR. Diving amongst the stars: NASA and its contributions to recreational scuba. *Tech Diving Mag*, Issue 3, June 2011.

5. Sheffield PJ, Vann RD, eds. *Flying After Recreational Diving Workshop Proceedings*. May 2, 2002. Durham, NC: Divers Alert Network; 2004.

6. Salama A. Accelerating no-fly time using surface oxygen. *Tech Diving Mag*, Issue 1, December 2010.

7. Pollock NW, Natoli MJ, Gerth WA, Thalmann ED, Vann RD. Risk of decompression sickness during exposure to high cabin altitude after diving. *Aviat Space Environ Med*. 2003 Nov; 74(11):1163-8. PMID: 14620473.

8. Robinson RR, Dervay JP, Conkin J. *An Evidenced-Based Approach for Estimating Decompression Sickness Risk in Aircraft Operations*. NASA/TM-1999-209374. Houston, Texas: NASA/Johnson Space Center; 1999.

9. Bason R, Yacavone DW. Loss of cabin pressurization in U.S. Naval aircraft: 1969-90. *Aviat Space Environ Med*. 1992 May; 63(5):341-5. PMID: 1599378.

10. Brooks CJ. Loss of cabin pressure in Canadian Forces transport aircraft, 1963-1984. Avi*at Space Environ Med*. 1987 Mar; 58(3):268-75. PMID: 3579812.

11. Powell MR. Diving amongst the stars: NASA and its contributions to recreational scuba. *Tech Diving Mag*, Issue 3, June 2011.

12. Hempleman HV. *The Unequal Rates of Uptake and Elimination of Tissue Nitrogen Gas in Diving Procedures*. Royal Naval Personnel Research Committee, RNP 62/1019, UPS 195, RNPL 5/60. London: Medical Research Council; 1960.

13. Hempleman HV. British decompression theory and practice. In: Bennett PB, Elliott DH, eds. *The Physiology and Medicine of Diving and Compressed Air Work. 1st Edition*. London: Bailliere Tindal & Cassell; 1969:290–318.

14. Berghage TE, Dyson CV, McCracken TM. Gas elimination during a single-stage decompression. *Aviat Space Env Med*. 1978; 49(10):1168–1172.

15. Powell MR, Spencer MP, Rogers RE. Doppler ultrasound monitoring of gas phase formation following decompression in repetitive dives. Seattle: Diving Science and Technology Corp.; 1988.

16. Powell MR. Oxygen and the diver. *Tech Diving Mag*, Issue 6, March 2012.

17. Weathersby PK, Hart BL, Flynn ET, Walker WF. Role of oxygen in the production of human decompression sickness. *J Appl Physiol*. 1987 Dec; 63(6):2380-7.

18. Powell MR. Diving amongst the stars: NASA and its contributions to recreational scuba. *Tech Diving Mag*, Issue 3, June 2011.

19. Torti SR, Billinger M, Schwerzmann M, Vogel R, Zbinden R, Windecker S, Seiler C. Risk of decompression illness among 230 divers in relation to the presence

and size of patent foramen ovale. *Eur Heart J.* 2004 Jun; 25(12):1014-20. doi: 10.1016/j.ehj.2004.04.028. PMID: 15191771.

20. Leffler CT. Effect of ambient temperature on the risk of decompression sickness in surface decompression divers. *Aviat. Space Environ. Med.* 2001; 72(5):477-483.

21. Gerth WA, Ruterbusch VL, Long ET. *The Influence of Thermal Exposure on Diver Susceptibility to Decompression Sickness.* TA03-09. NEDU TR 06-07. Panama City, FL: Navy Experimental Diving Unit; 2007.

22. Blatteau JE, Gempp E, Balestra C, Mets T, Germonpre P. Predive sauna and venous gas bubbles upon decompression from 400 kPa. *Aviat Space Environ Med.* 2008 Dec; 79(12):1100-5.

23. Gempp E, Blatteau JE, Pontier JM, Balestra C, Louge P. Preventive effect of pre-dive hydration on bubble formation in divers. *Br J Sports Med.* 2009; 43:224-228. doi:10.1136/bjsm.2007.043240.

24. Wilbur JC, Phillips SD, Donoghue TG, Alvarenga DL, Knaus DA, Magari PJ, Buckey JC. Signals consistent with microbubbles detected in legs of normal human subjects after exercise. *J Appl Physiol.* 2010; 108:240–244. doi:10.1152/japplphysiol.00615.2009.

25. Powell MR. Diving amongst the stars: NASA and its contributions to recreational scuba. *Tech Diving Mag,* Issue 3, June 2011.

26. Nishi RY, Jankowski LW, Tikuisis P. Effect of exercise on bubble activity during diving. In: *Operational Medical Issues in Hypo- and Hyperbaric Conditions.* Papers presented at the RTO Human Factors and Medicine Panel Symposium; Toronto, Canada; 16-19 October 2000. Neuilly-sur-Seine, France: NATO Research and Technology Organization; 2001.

27. U.S. Department of the Navy. *U.S. Navy Diving Manual, Revision 6.* NAVSEA 0910-LP-106-0957. Washington, DC: Naval Sea Systems Command; 2008.

28. Lang MA, Egstrom GH, eds. *Proceedings of the AAUS Biomechanics of Safe Ascents Workshop.* September 25-27, 1989. Woods Hole, Mass. Costa Mesa, CA: American Academy of Underwater Sciences; 1990.

29. Blatteau JE, Pontier JM. Effect of in-water recompression with oxygen to 6 msw versus normobaric oxygen breathing on bubble formation in divers. *Eur J Appl Physiol.* 2009 Jul; 106(5):691-5. doi: 10.1007/s00421-009-1065-y. PMID: 19424716.

30. Smith CR. Intravenous administration of perfluorocarbon emulsions as a non-recompression therapy for decompression sickness [dissertation]. Richmond, VA: Virginia Commonwealth University, School of Medicine; June 2008.

31. Suunto Fused™ RGBM. Suunto Oy; 2012. Available at: http://ns.suunto.com/pdf/Suunto_Dive_Fused_RGBM_brochure_EN.pdf.

32. Walder DN. Adaptation to decompression sickness in caisson work. In: Tromp SW, Weihe WH, eds. *Biometeorology: Proceedings of the Third International Biometeorological Congress*; 1963 Sep 1-7; Pau, France. Oxford: Pergamon Press; 1967:350-9.

33. Gilliam B. Evaluation of decompression sickness incidence in multi-day repetitive diving for 77,680 sport dives. *SPUMS J*. 1992; 22(1):24-30.

34. Vann RD, Mitchell SJ, Denoble PJ, Anthony TG, eds. *Technical Diving Conference Proceedings*; 2008 January 18-19. Durham, NC: Divers Alert Network; 2009.

35. Zanchi J, Ljubkovic M, Denoble PJ, Dujic Z, Ranapurwala SI, Pollock NW. Influence of repeated daily diving on decompression stress. *Int J Sports Med*. 2014 Jun; 35(6):465-8. doi: 10.1055/s-0033-1334968. Epub 2013 Jun 14. PMID: 23771833.

36. Vann RD, Mitchell SJ, Denoble PJ, Anthony TG, eds. *Technical Diving Conference Proceedings*; 2008 January 18-19. Durham, NC: Divers Alert Network; 2009.

37. Bennett PB, Wienke B, Mitchell S, eds. *Decompression and the Deep Stop Workshop Proceedings*; 2008 June 24-25; Salt Lake City, Utah. Durham, NC: Undersea and Hyperbaric Medical Society; 2008.

38. Spisni E, Marabotti C, De Fazio L, Valerii MC, Cavazza E, Brambilla S, Hoxha K, L'Abbate A, Longobardi P. A comparative evaluation of two decompression procedures for technical diving using inflammatory responses: compartmental versus ratio deco. *Diving Hyperb Med* 2017, 47 (1): 9-16.

39. Madden LA, Laden G. Gas bubbles may not be the underlying cause of decompression illness: The at-depth endothelial dysfunction hypothesis. *Med Hypotheses*. 2009 Apr; 72(4):389-92. doi: 10.1016/j.mehy.2008.11.022. PMID: 19128890.

40. Xu W, Zhang K, Wang M, Wang H, Liu Y, Buzzacott P. Time course of endothelial dysfunction induced by decompression bubbles in rats. *Front. Physiol*. doi: 10.3389/fphys.2017.00181.

41. Zhang K, Jiang Z, Ning X, Yu X, Xu J, Buzzacott P, Xu W. Endothelia-targeting protection by escin in decompression sickness rats. *Scientific Reports 7*, Article number: 41288 (2017). doi:10.1038/srep41288.

42. Dujić Ž, Duplančic D, Marinovic-Terzić I, Baković D, Ivančev V, Valic Z, Eterović D, Petri NM, Wisløff U, Brubakk AO. Aerobic exercise before diving reduces venous gas bubble formation in humans. *J Physiol*. 2004 March 16; 555(Pt 3):637–642.

43. Wisløff U, Richardson RS, Brubakk AO. Exercise and nitric oxide prevent bubble formation: a novel approach to the prevention of decompression sickness? *J Physiol*. 2004 March 16; 555(Pt 3):825–829.

44. Wisløff U, Richardson RS, Brubakk AO. NOS inhibition increases bubble formation and reduces survival in sedentary but not exercised rats. *J Physiol.* 2003 January 15; 546:577-582. doi:10.1113/jphysiol.2002.030338.

45. Doolette DJ, Upton RN, Grant C. Altering blood flow does not reveal differences between nitrogen and helium kinetics in brain or in skeletal miracle in sheep. *J Appl Physiol.* 118: 586–594, 2015. First published December 18, 2014; doi:10.1152/japplphysiol.00944.2014.

46. Doolette DJ, Gault KA, Gerth WA. Decompression from He-N_2-O_2 (trimix) bounce dives is not more efficient than from He-O_2 (heliox) bounce dives. *NEDU TR* 15-04, 2015.

Appendix A: Using Hydrogen as a Diving Gas

1. Gauch F, Greenfire M, eds. Comex keeps up the high pressure. *Comex Magazine*; Issue 4, January 2009.

2. Fructus XR. Hydrogen, pressure and HPNS. In: Brauer R, ed. *Hydrogen as a Diving Gas: Proceedings of the Thirty-Third Undersea and Hyperbaric Medical Society Workshop*; February 1987; Institute for Marine Biomedical Research, Wilmington, N.C. Bethesda, MD: Undersea and Hyperbaric Medical Society; 1987.

3. Smith EB. Hydrogen: theoretical considerations. In: Brauer R, ed. *Hydrogen as a Diving Gas: Proceedings of the Thirty-Third Undersea and Hyperbaric Medical Society Workshop*; February 1987; Institute for Marine Biomedical Research, Wilmington, N.C. Bethesda, MD: Undersea and Hyperbaric Medical Society; 1987.

4. Fructus XR. Hydrogen narcosis in man. In: Brauer R, ed. *Hydrogen as a Diving Gas: Proceedings of the Thirty-Third Undersea and Hyperbaric Medical Society Workshop*; February 1987; Institute for Marine Biomedical Research, Wilmington, N.C. Bethesda, MD: Undersea and Hyperbaric Medical Society; 1987.

5. Fife WP. The toxic effects of hydrogen-oxygen breathing mixtures. In: Brauer R, ed. *Hydrogen as a Diving Gas: Proceedings of the Thirty-Third Undersea and Hyperbaric Medical Society Workshop*; February 1987; Institute for Marine Biomedical Research, Wilmington, N.C. Bethesda, MD: Undersea and Hyperbaric Medical Society; 1987.

6. Kayar SR, Fahlman A. Decompression sickness risk reduced by native intestinal flora in pigs after H2 dives. *Undersea Hyperb Med.* 2001; 28(2):89–97. PMID: 11908700.

7. Fahlman A, Kayar SR, Himm J, Tikuisis P. Probabilistic modeling of decompression sickness (DCS) risk for pigs in H2. UHMS Annual Meeting, Stockholm, Sweden; 2000.

Appendix B: Calculating the acceleration in post-diving no-fly time associated with breathing surface oxygen

1. Salama A. Accelerating no-fly time using surface oxygen. *Tech Diving Mag*, Issue 1, December 2010.

2. Sheffield PJ, Vann RD, eds. *Flying after recreational diving workshop proceedings.* May 2, 2002. Durham, NC: Divers Alert Network; 2004.

3. Bühlmann AA. Tauchmedizin. Berlin: *Springer-Verlag*; 1995.

4. Boycott AE, Damant GCC, Haldane JS. The prevention of compressed air illness. *J Hyg* (London). Jun 1908; 8(3): 342–443.

INDEX

decompression tables, 1, 8, 10, 19-20, 27, 65, 67, 86
  NOAA-Hamilton Trimix Decompression Tables,
    65
  U.S. Navy Standard Air Decompression Tables,
    10, 20
  See also dive tables
decompression zone, 26-27, 88
  multilevel ascents and, 88
  trimming, 26-27
deep stops, 23-27, 49, 58-59, 69, 83, 89, 96-98, 100,
  102-103
  DCS risk, 30, 63-64, 77, 86, 90, 107
  French Navy study, 97
  gas volume expansion (GVE) technique, 25-26,
    83, 96
  NEDU study, 97
  postdive fatigue and, 23, 30-31
  Pyle stops, 23-26, 83, 96
  thermodynamic theory, 49
  VPM, 52
Defence and Civil Institute of Environmental
  Medicine, 11, 62
  DCIEM 1983 tables, 62
  DCIEM manual, 62
  dive tables, 11, 62
  Kidd-Stubbs model, 62
Defence Research and Development Canada
  (DRDC), 11. *See* Defence and Civil Institute
  of Environmental Medicine (DCIEM)
dehydration, 23, 66, 90-91, 93, 101
demand mask, 75
desaturation, 16, 19, 36, 39, 74, 104
descend/descent, 14-15, 30, 53, 67, 88, 106
  multilevel, 88
  rate, 53-68, 106
diffusion, 10, 15-16, 18-19, 33, 43, 45, 52, 59, 61-62,
  69-71, 81
  gradients, 18, 33-34, 43, 59, 81
  limited, 8, 19, 22, 28, 47, 59, 61, 69, 81, 84, 87,
    91, 100
  slab, 10, 61-62
  transdermal, 43
diffusivity ratio, 39, 42, 101, 105
dissolved-gas (Haldanean) models, 11, 24, 46, 49-50,
  52, 59, 63, 67, 84, 89, 96, 98
  adaptive algorithms, 22
  bubble-reduction factors, 58
  Bühlmann, 11, 24, 50, 59, 67, 84, 98
  flawed, 46
  gradient factors, 25-27
  Haldane, 11, 49-50, 52, 63, 67, 84
  Haldanean/neo-Haldanean models, 49-50, 52,
    63, 67
  perfusion limited, 19
  Workman, 20-21
  ZH-L models, 21-24, 26-27, 32, 39, 50, 53-56,
    58-59, 64, 67-68, 79-81, 83-84, 90, 98, 100

distribution ratio, 36
dive computers, 3, 11, 22-23, 28, 58-59, 61, 63, 74,
  82, 85, 95, 98
  Cochran Undersea Technology, 11, 63
  K-S PADC, 61-62
  pneumatic, 61-62
  Suunto, 5, 59, 95
  Uwatec, 3, 5, 11, 29, 59
  USN E-L algorithm, 11, 63
dive profiles, 58, 63, 74, 103
  analyzers, 64
  linear, 68
  square, 74
Divers Alert Network, 1-2, 28, 73-74, 77-78, 82,
  95-96
  Technical Diving Conference, 95-96
dive schedules. *See* decompression schedules
dives, experimental, 10, 67
dive tables, 6, 64, 74, 79, 98
  AB, 1, 3-16, 18-41, 43-50, 52-59, 61-101,
    103-111
  altitude, 5, 7-8, 11, 21, 40, 47, 72-74, 77-78, 80,
    82-83, 87, 92-93, 101, 108, 110
  British Sub-Aqua Club (BSAC), 11, 29
  C&R (Bureau of Construction and Repair), 10
  Comex, 67, 104, 107
  compressed air, 9, 19, 35, 66
  DCIEM, 62
  deep diving, 8, 36, 45
  Gerth and Doolette, 63
  Haldanean (neo-Haldanean), 10, 48-50, 52, 58,
    63, 67
  Navy Exceptional Exposure Tables, 3
  NOAA-Hamilton Trimix Decompression Tables,
    65
  Tables du Ministère du Travail 1992 (MT92), 67
  U.S. Navy, 61-63, 74, 77-78, 97
  USN E-L, 11, 62-63, 65, 96
  Yarborough, 10
  See also decompression tables
Diving Equipment and Manufacturing Association
  (DEMA), 23, 29, 75, 79
diving gases. *See* gases
Donald, Kenneth W., 85
Doolette, David, 11, 63, 96
Doppler technology, 11, 47
  precordial, 47, 87, 90
  transcranial (TCD), 87
dual-phase models, 48-49, 66, 68, 89, 96, 98
  bubble models, 48
  AB model, 66-67
  asymptomatic (silent) bubbles, 24, 46-48, 53, 96
  Brian Hills, 48
  critical volume algorithm, 53, 89
  evolution of, 48
  phase equilibrium, 49
  reduced gradient bubble model (RGBM), 58-59

thermodynamic and kinetic approach, 48
varying permeability model (VPM), 11, 49
  with Boyle's Law compensation (VPM-B), 54
  VPM-B conservatism, 55
  VPM-B variations, 56
Duke University's Center for Hyperbaric Medicine
  and Environmental Physiology, 78

enriched air nitrox (EAN), 30. *See* nitrox
equilibrium, 14, 16, 18, 34, 40, 49-51, 82, 84, 96
equivalent air depth (EAD), 31, 38
equivalent narcotic depth (END), 38, 103
European Association of Technical Divers (EATD),
  29
exercise, 23, 47, 80, 91-92, 96, 98, 100-101
  DCS risk and, 90
  during decompression, 50, 63, 68, 73, 90, 92
Exley, Sheck, 3, 28, 37
exponential decay, law of, 15
exponential function, 8, 15-16
exponential-linear kinetics, 64
exponential rate, 11, 63

fatigue, postdive, 23, 30-31
flow-mediated dilation (FMD), 31
flushing gas, 106-107
flying after diving, 73, 108
  accelerating no-fly time, 74
  calibration study, 82
  consensus guidelines, 73
  preflight surface intervals, 73
  repetitive flights, 78
  surface oxygen and, 79
  U.S. Navy flying-after-diving table, 77
French Navy, 67, 97

Gardette, Bernard, 97
gas elimination, 72, 90, 92
gases, 8, 10, 13-15, 20, 22, 29-31, 36-37, 39, 41,
  43-46, 50-51, 61, 69, 85-86, 94-96, 100, 104-105,
  108-109
  air, 3, 5-7, 9-10, 13-14, 16, 18-23, 29-35, 37-38,
    40-43, 47-48, 50, 63-64, 66-68, 70, 72-75,
    78-81, 84-85, 89, 93-95, 97-98, 100, 106-110
  argon, 13, 29, 36
  argox, 36
  breathing, 68, 97-98
  carbon dioxide, 21, 50, 73, 98
  EAN. *See* nitrox
  flushing, 106-107
  heliair, 37
  heliox, 35, 37, 67, 75, 101, 104, 107
  helium, 11, 20, 22, 35-45, 50, 53-55, 63, 68-69,
    79, 81, 83, 96, 98-101, 104-105, 107-109
  hydreliox (hydroheliox), 107
  hydrogen, 36, 104-105, 107

hydrox, 105-107
mixed, 20, 35-36, 104
narcotic effects of, 38
neon, 36, 65
nitric oxide, 101
nitrogen, 6-8, 10-11, 13-16, 18-20, 22, 29-33,
  35-45, 50, 52-55, 63, 65, 68-69, 73, 76, 79,
  81-85, 96, 99, 101, 104-109
nitrox, 2-3, 28-32, 34-35, 38, 41, 43, 54, 67, 70,
  81, 108-109
nonmetabolic, 43, 45
oxygen, 3, 5-8, 10, 13-14, 20, 22, 29-32, 34-38,
  40, 43, 45, 47, 49-50, 59, 63-65, 68, 72-82,
  84-86, 92-95, 98, 100-102, 104-110
oxygen enriched air (OEA), 30
steady state, 72
trimix, 37-38, 40-41, 44-45, 54, 64-65, 75, 79-81,
  88, 97-99, 101, 103-104, 107-110
vascular, 31-32, 68, 98, 100
See also inert gas; solubility
gas exchange, 20, 41, 43, 59, 72-73, 91
gas, inert. *See* inert gas
gas laws, 7, 11, 14-15, 18, 22, 30, 39, 52, 54, 59
  Boyle's Law, 7, 11, 14, 52, 54, 59
  Dalton's Law, 14
  Graham's Law, 22, 39
  Henry's Law, 15, 30
gas switches, 3, 38, 43, 88
  multilevel descents, 88
gas transport, 15-16
gas uptake, 10, 18-20, 36, 50, 52, 61, 72, 84, 88, 90,
  97
gas volume, 14, 25, 47, 53, 56, 100
  critical volume, 53-54, 89
gas volume expansion (GVE), 25-26, 83, 96
Gault, Keith, 96
Gerth, Wayne, 11, 63, 96
Gilliam, Bret, 3, 28, 37, 74, 95
  acclimatization studies, 95
  IAND, 28
  IANTD, 2, 28-29
  NAUI, 2, 28-29
  TDI, 23, 29-31, 74-76, 79, 81, 93, 100, 103
  Uwatec, 3, 5, 11, 29, 59
Gleason, Bill, 28
Goldman, Saul, 69-70
gradient, 16, 18, 25-27, 32-34, 41, 43, 52, 54-56, 58-59,
  64, 66-67, 79-81, 83, 87, 89, 95-96, 98-100, 103
  diffusion, 33
  inert gas, 33
  offgassing, 81
  partial pressure, 43
  reduced, 58, 95, 100
  reversals, 87
  supersaturation, 52-54, 59, 65, 89
gradient factors (GF), 25-27, 32, 54-55, 58-59, 64,
  67, 80, 83, 96, 98, 100

Gurr, Kevin 29

Hahn, Max, 23
Haldane equation, 19, 50, 57, 80-81
Haldane, John Scott, 8-12, 15-20, 27, 48-50, 52, 57-58, 61-63, 65-67, 74, 78, 80-81, 84-85, 109
    algorithm, 8, 11, 27, 50, 57, 62-63, 74, 109
    ascent rates, 17-18
    dissolved-gas models, 18
    dive/decompression tables, 8, 10, 19-20, 27, 65, 67, 74
    Father of Modern Decompression Theory, 15
    goat studies, 9-10, 19, 84-85
    Haldane equation, 19, 50, 57, 80-81
    halftimes, 8, 10-11, 15, 18-20
    Journal of Hygiene, 8
    points, 9
    pressure reduction ratio, 19-20
    stage decompression, 8
Haldanean/neo-Haldanean models, 11, 49-50, 52, 63, 67
    linear pressure reduction ratios (M-values), 50
    with RGBM, 67
halftimes, 8, 10-11, 15, 18-20, 22-23, 39, 41-42, 47, 50-51, 59, 61-62, 65, 67, 84-85, 101, 105
    ultralong, 18, 84
    varying, 8, 11, 15, 18-19, 50-51, 59, 67
    See also compartments
Hamilton, Bill, 3, 65, 71
Hamilton-Kenyon model. *See* Tonawanda IIa
Hamilton Research Ltd., 65
Hansen, Raymond, 10, 20
Harvey, Edmond Newton, 46
Hawkins, James, 10, 20
heat stress, 92
heliair, 37
heliox, 35, 37, 67, 75, 101, 104, 107
helium, 11, 20, 22, 35-45, 50, 53-55, 63, 68-69, 79, 81, 83, 96, 98-101, 104-105, 107-109
    compared with nitrogen, 104-105
    halftimes, 41-42
    HPNS and, 37-38
    narcotic effect of, 37
    solubility, 43-45
    speed of sound in, 37
    trimix, 37-38, 40-41, 79-81
Hemingway, Ross, 12, 56, 84
    /E, 12, 56, 84
    VPM-B/E, 56
hemoglobin, 72
Hemphill, Brett, 76
Hempleman, Henry Valance "Val", 10, 12, 53, 61, 71, 85
Hennessy, Tom, 53
Henry's Law, 15
high-pressure nervous syndrome (HPNS), 37-38, 104

Hill, Sir Leonard, 19
Hills, Brian, 48-49, 67
Hobbs, Gene, 77
Hoffman, Don, 11-12, 53
Hooke, Robert, 7
Hugon, Michel, 97
hydration, 23, 66, 76-77, 90-93, 101
Hydra X project, 104
hydreliox (hydroheliox), 107
hydrogen, 36, 104-105, 107
    isobaric counterdiffusion, 43
    ICD, 4, 43-45, 64-65, 99, 105, 107
    narcotic potency, 37-38, 104, 107
    thermal conductivity, 37, 104
hydrox, 105-107
hyperbaric chamber, 9, 62, 85, 90, 96, 100, 104. *See also* recompression chamber
hypoxia, 32, 36

Imbert, Jean Pierre, 67, 71
inert gas, 10-11, 13, 18-23, 33-34, 36-37, 39-41, 43, 46, 49-53, 59, 61-63, 67, 69-74, 80-81, 83-86, 90, 92, 96, 98, 103-104, 107, 109
    alveolar, 21, 40-41
    arterial, 67
    exchange, 41, 43
    gradients, 33-34
    lipid solubility, 36-37
    partial pressure of, 52
    physical properties of inert gas, 37
        molecular weight, 37
        solubility in blood, 36
        relative narcotic potency (RNP), 37-38
    residual, 80, 96
    saturation and desaturation speeds, 104
    steady state, 72
    substitutes, 36
    tension, 43, 67
    uptake and elimination, 50, 61-63, 84-85, 88, 103, 109
inherent unsaturation, 72
initial critical radius, 53, 105
inner ear, 4, 44-45
    DCS, 44
    ICD, 44-45
interconnected, 59, 62, 69-70
    in parallel, 69
in-water recompression, 94, 102
International Association of Nitrox and Technical Diving (IANTD), 2, 28-29
International Association of Nitrox Divers (IAND), 28
isobaric counterdiffusion (ICD), 4, 43-45, 64-65, 99, 105, 107
Kelley, Patrick, 20
Kenyon, David, 65, 71

no-stop dives, 17, 24, 58, 86, 93
novel approaches, 100, 102
nucleation, 46-48, 50
    stress assisted nucleation, 46-48

offgassing, 14-15, 18, 33-34, 36, 41, 58, 63, 70, 76, 81-82, 85, 90-92
omitted decompression, 92-94, 102
ongassing, 14-15, 18, 25, 30, 33-34, 41, 51, 70, 77, 81, 85, 90
optimization, 68
overpressure ratio, 20
oxygen (O_2), 3, 5-8, 10, 13-14, 20, 22, 29-32, 34-38, 40, 43, 45, 47, 49-50, 59, 63-65, 68, 72-82, 84-86, 92-95, 98, 100-102, 104-110
    argox, 36
    bends, 7-8, 43-44, 50, 75-76, 85, 92
    bound to hemoglobin, 72
    conversion to carbon dioxide, 72
    hypoxia, 32, 36
    narcotic effects of, 38, 104
    oxygen enriched air, 30
    oxygen window, 49, 72-73, 102
    partial pressure of, 13, 15, 20, 30-31, 33-34, 36, 38, 52
    poisoning, 43
    prebreathe, 47, 78
    pure, 3, 8, 48, 63-64, 74-76, 81, 93-95, 109-110
    saturation therapy, 75
    solubility of, 72
    surface, 74-75, 79, 81-82, 94, 108
    toxicity, 7, 30, 32, 34-35
oxygen bends, 50, 85. *See* bends; decompression sickness
oxygen enriched air (OEA), 30
oxygen toxicity, 7, 30, 32, 34-35
    acute CNS, 32
    central nervous system, 7
    chronic pulmonary, 32
    Lorrain Smith effect, 32
oxygen window, 49, 72-73, 102

PADI, 28-29
Palmer, Rob, 29
partial pressure, 9, 13-15, 20-22, 30-31, 33-36, 38-41, 43, 49, 52, 72, 105, 109
    alveolar, 16, 33-34, 36, 81-82
    gradients, 43
    inspired, 16, 33-34, 36
    of carbon dioxide, 72
    of inert gases, 52
    of nitrogen, 20, 30-31, 38, 52
    partial pressure of oxygen, 13, 30
    oxygen window, 49, 72
    vacancy, 72
partition coefficient/constant/ratio, 36
patent foramen ovale (PFO), 44, 66, 86-88, 101

pearl divers, 48, 67
perfluorocarbons, 95
perfusion, 15, 18-19, 22-23, 49, 59, 61, 68-69, 71-72, 81, 84, 100
    blood perfusion, 22, 68, 72
    intermittent perfusion, 49
    limited perfusion, 19, 59, 61, 69, 81, 84
permeability, 11, 49-50, 52
phase change, 46
pneumatic computer, 62
Pollock, Neal, 78
Powell, Mark, 75
Powell, Michael, 7, 46, 80
    scale for grading Doppler-detected bubbles, 80
pressure, 5, 7, 9-10, 13-16, 18-23, 25-27, 30-31, 33-41, 43-45, 48-53, 57-58, 61, 67-68, 72-73, 75, 78, 81-83, 85-87, 89, 92, 96, 100, 103-105, 109
    absolute pressure, 10, 14, 16, 19, 21, 25
    alveolar pressure, 16, 21, 33-34, 36, 40-41, 57, 81
    ambient pressure, 7, 14, 16, 18, 20-21, 26-27, 30-31, 33-34, 36, 43, 48, 51-52, 61, 68, 82, 96
    atmosphere (atm), 13, 51
    atmosphere absolute (ATA), 13
    atmospheric (bar), 13-14, 38, 40, 73
    barometric, 13, 82. *See also* atmospheric pressure
    breathing mix, 37-38
    bubble, 50-51, 53
    cabin, 75, 78
    crushing, 52-53, 89
    depth, 21
    elevated, 15
    gradients, 33-34
    Laplace, 51
    maximum allowed, 51
    overpressure ratios, 20
    partial, 9, 13-15, 20-22, 30-31, 33-36, 38-41, 43, 49, 52, 72, 105, 109
    pressure reduction ratio, 16, 19-20, 50
    saturation, 44, 53, 105
    sea level, 73
    surface tension, 50-51
    water vapor, 40
pressure gradients, 33-34
pressure-reversal effect, 37
probabilistic decompression models, 63, 77
pulmonary dead space, 72
Pyle, Richard, 23-26, 83, 96
Pyle stops, 24

rebreathers, 29, 58. *See also* closed-circuit rebreathers
recompression, 28, 30, 53, 67, 88-90, 93-94, 102
    in-water, 94, 102
    multilevel (sawtooth) profiles, 53, 67, 88
recompression chamber, 93-94. *See also* hyperbaric chamber
repetitive dives, 26, 48, 54, 58, 62-63, 80-81, 95-96

respiratory quotient, 21
reverse dive profiles, 58
Reymenants, Ben, 76, 84
Royal Naval Physiological Laboratory (RNPL), 10, 54, 61
Royal Navy, 10, 15-16, 61
　Royal Navy dive tables, 10, 61
rule of fifth, 44-45, 105
Russian studies, 83
Rutkowski, Dick, 28-29

safe ascent depth (SAD), 62, 98, 104
safety stops, 24. *See also* decompression stops
saline intravenous therapy, 95
Salm, Albrecht, 23
saturation, 9-10, 15-19, 21-22, 33, 36, 39-40, 43-44, 46, 48-49, 52-54, 59, 62, 65, 68, 72, 74-75, 82, 84, 89, 96, 104-105
　speed saturation, 39, 104
SAUL model, 69-70
　interconnected bubble model (ICBM), 70
　interconnected model (ICM), 70
Schreiner equation, 50, 57
Schreiner, Heinz, 20-21, 41, 50, 57, 65
　algorithm, 65
Scubapro Decompression Meter, 74
sea level, 7, 13, 21, 29, 36, 40, 47, 73, 80-82, 96, 103
Sharm El Sheikh, Egypt, 5, 31
Shaw, Dave, 56
Shearwater Research, 12
　/GFS, 12, 56, 59, 84
　VPM-B, 54-60
Shilling, Charles, 10, 20
shunting, 86
Skiles, Wes, 3
skin bends, 43, 76. *See also* decompression sickness
Skin Diver magazine, 28
Smith, E.B., 32, 104
software, 3-4, 6, 11-12, 22, 24, 26, 50, 54, 63-65, 75-76, 80-84, 88-89, 98
solubility, 15, 18, 22, 32, 36-37, 39-40, 43-45, 72, 90, 92, 105
　coefficient, 15, 22, 37, 39, 44-45, 105
　in blood, 36, 39, 105
　lipid, 36-37
SSI, 28
statistical inference model, 63, 71
steady state, 72
Stillson, George, 10
stops, 3, 16-18, 23-27, 36, 41, 44, 49-50, 53-54, 58-59, 63-64, 69, 73, 75, 83, 88-89, 93-94, 96-98, 100-103, 106-110
　deep, 23-27, 49, 58-59, 69, 83, 89, 96-98, 100, 102-103
　extended, 73

gas switches, 3, 38, 43, 88
no stops, 17, 24, 58
Pyle stops, 24
safety, 17, 24
See also decompression stops
Stubbs, Roy, 61-62, 65, 69-71
Sub-Aqua Association (SAA), 29
subsaturation dives, 53
supersaturation, 9-10, 16-18, 21, 43, 46, 48-49, 52-54, 59, 62, 65, 68, 89, 105
　critical, 16-17, 43
　gradients, 18, 54, 59, 89
　inherent unsaturation, 72
　initial allowable, 54
　maximum allowed, 52
　nil, 49
　oxygen window, 49
　ratio, 10, 62
　zero, 49
surface intervals, 28, 30, 58, 73, 75, 80, 91, 101
　oxygen breathing and, 74, 77-80
　preflight, 73-74, 78-79, 82
　See also flying after diving
surface oxygen, 74-75, 79, 81-82, 94, 108
surface tension, 50-51. *See also* Laplace pressure
surfactant, 46, 58
Suunto RGBM, 59, 95

Tables du Ministère du Travail 1992 (MT92), 67
tacky adhesion, 46
Taylor, Glenn H., 28, 78
Tech Diving Mag, 4, 82
technical diving, 4, 28-29, 95-96
Technical Diving International (TDI), 23, 29-31, 74-76, 79, 81, 93, 100, 103
Tektite saturation dives, 54
temperature, 14-15, 22-23, 29, 66, 73, 76, 90, 92, 101, 103
Thalmann, Edward, 11-12, 63-64, 71, 97
　probabilistic decompression model validation, 63-64, 77
　USN E-L algorithm, 11, 63
　VVal Thalmann algorithm, 63, 97
thermal status. *See* temperature
thermodynamic theory, 49
　thermodynamic and kinetic approach, 48
tiny bubble group, 49-50
tissues, 7-9, 11, 14-16, 18-19, 25, 30, 33, 36, 39-41, 43-46, 49-51, 59, 61-64, 66, 69-72, 76, 80-82, 84-85, 90-92, 96, 104, 107
　desaturation, 16
　gas transfer in, 19
　halftimes, 18-19, 61, 84
　interconnectivity, 69
　loadings, 40-41, 55, 81, 83, 96, 99
　micronuclei of tissues, 91

microregions of tissues, 49
production of carbon dioxide, 72
temperature effects, 90, 92
tension, 16, 26, 34, 43-45, 50-51, 53, 80-82
types of tissues, 49
See also compartments
Tonawanda IIa (Hamilton-Kenyon), 65
Haldane-Workman-Schreiner algorithm, 65
11F6, 65
MF11F6, 65
MM11F6, 65
toxicity, oxygen 7, 30, 32, 34-35
chronic pulmonary, 32
CNS, 7, 32, 34
transport medium, 22, 39, 105
tribonucleation, 46
trimix, 37-38, 40-41, 44-45, 54, 64-65, 75, 79-81, 88,
97-99, 101, 103-104, 107-110
Turner, Parker, 3

Ultimate Planner, 4, 6, 21, 23, 45, 55, 65, 73, 80-81,
84-85, 90, 99-100, 105
Dec compartment set, 84
Schreiner's value, 21
/U, 21, 84-85, 90, 100
ZH-L16D, 23, 90
undeserved/unearned hit, 6, 67, 86, 94
U.S. Air Force, 78
U.S. Navy, 3, 10-11, 18, 20-21, 35, 54, 57, 61-63, 70,
74, 77-78, 93-94, 97
datasets, 63
deep stops, 94
dive tables, 64, 74, 79, 98
flying after diving, 73, 108
USN E-L model, 11, 62-63, 65, 96
U.S. Navy Standard Air Decompression
Tables, 10, 20
U.S. Navy in-water recompression protocol, 93
Navy Experimental Diving Unit (NEDU), 11, 57,
63, 90, 96-97
omitted decompression protocols, 94
U.S. Navy exponential linear (USN E-L) model, 11,
62-63, 65, 96. *See also* VVal Thalmann
U.S. Navy Diving Manual, 18
USS *F-4*, 10
USS *Monitor*, 65
U.S. Special Operations Forces, 78
USS *Squalus*, 35
Uwatec, 3, 5, 11, 29, 59

Valsalva maneuvers, 87
varying gradient model (VGM), 59
varying permeability model (VPM), 4, 11, 24, 49-50,
52-60, 64-65, 75, 79-80, 83-85, 89, 105, 107,
109-110
adjusted allowable supersaturation gradients, 54

bubble seeds, 50, 84
comparison with ZH-L16, 24, 50, 53-56, 58-59,
64, 79-80, 83-84, 109-110
crushing pressure, 52-53, 89
DCAP, 65
initial allowable supersaturation gradients, 54
VPM-0, 54, 59
VPM-A, 54
VPM-values, 53
VPM with Boyle's Law compensation (VPM-B),
4, 11, 54-60, 64, 80, 83-85, 89, 105, 107,
109-110
comparison with ZH-L16B, 54-56, 58-59, 64,
80, 83-84, 109-110
multilevel dives, 88
ultralong halftimes, 84
VPM-B conservatism, 55, 64, 109-110
VPM-B variations, 56
vasoconstriction, 23, 76, 90, 100
vasodilation, 90, 92
venous blood, 16, 72, 86
venous gas bubble grade, 31
venous gas emboli (VGE), 17, 98
vestibular bends, 44
viscous adhesion, 46
visual analog scale, 31
VVal Thalmann, 63, 97

washout treatment, 96, 101
water vapor, 16, 21, 33, 40, 50
Wienke, Bruce, 58-59, 77, 80
Workman, Robert, 11-12, 20-21, 27, 65
algorithm, 65
computer programming, 20
conversion from Bühlmann's M-values, 21
helium, 20
linear projection, 20
M-values, 20-22
pressure reduction ratio, 19-20

Yarborough, O.D., 10
Yount, David, 11-12, 24, 49, 53-54, 58
tiny bubble group, 49-50

Zetterström, Arne, 107
ZH-L models (Bühlmann), 21-24, 26-27, 32, 39, 50,
53-56, 58-59, 64, 67-68, 79-81, 83-84, 90, 98,
100, 109-110
ZH-L8 ADT, 22, 59
ZH-L12, 21-22, 27
ZH-L16, 22-24, 27, 32, 39, 50, 54-55, 58-59, 64,
67, 79-81, 83-84, 90, 98, 100, 109-110
comparison with AB-2, 67
comparison with VPM-B, 54-57

CPSIA information can be obtained
at www.ICGtesting.com
Printed in the USA
BVHW012056150821
614245BV00002B/5